Teacher's Manual

THE EARTH AND AGRISCIENCE

JOHN R. CRUNKILTON

Agricultural Educator
Virginia Polytechnic Institute
and State University

AgriScience and Technology Series

Series Editor
Jasper S. Lee, Ph.D.

SUSAN L. OSBORNE

Agricultural Educator
Urbana, Illinois

MICHAEL E. NEWMAN

Agricultural Educator
Mississippi State University

EDWARD W. OSBORNE

Agricultural Educator
University of Illinois

JASPER S. LEE

Agricultural Educator
Mississippi State, Mississippi

Interstate Publishers, Inc.
Danville, Illinois

THE EARTH AND AGRISCIENCE TEACHER'S MANUAL.
COPYRIGHT © 1995 BY INTERSTATE PUBLISHERS, INC. All rights reserved. Printed in the United States of America.

ISBN 0-8134-3016-X

1 2 3
4 5 6
7 8 9

Order from

Interstate Publishers, Inc.

510 North Vermilion Street
P.O. Box 50
Danville, IL 61834-0050

Phone: (800) 843-4774
Fax: (217) 446-9706

CONTENTS

To the Teacher 5

Overview of The Earth and AgriScience 6

Chapter 1 What Our Earth Provides 9

Chapter 2 How We Use the Earth's Resources 11

Chapter 3 Sustaining the Earth's Resources 13

Chapter 4 What We Need 15

Chapter 5 Meeting Our Needs 17

Chapter 6 Beyond Food: Ornamental Horticulture 19

Chapter 7 Using Forest Resources 21

Chapter 8 Getting Products to Us 23

Chapter 9 Using Science in Agriculture 25

Chapter 10 Sustaining Ecosystems 27

Chapter 11 Growing Plants 30

Chapter 12 Raising Animals 32

Chapter 13 Keeping Animals As Pets 34

Chapter 14 Using Power Machinery 36

Chapter 15 Helping Plants and Animals Grow 38

Chapter 16 Keeping a Good Environment 40

Chapter 17 Keeping Our Soil 42

Chapter 18	Keeping Our Water	44
Chapter 19	Keeping Our Air	46
Chapter 20	Keeping Our Wildlife	48
Chapter 21	Preventing Pollution	50
Chapter 22	Recycling	52
Chapter 23	Careers in AgriScience	54
Chapter 24	Getting the Education	56
Chapter 25	Developing Personal Skills	58
Chapter 26	Being a Good Citizen	60
Chapter 27	Keeping Healthy	62

TO THE TEACHER

The Earth and AgriScience uses an exciting approach to introduce students to agriculture and environmental science. The stress is on properly using the natural resources that are available. How people use the resources to meet their needs is an important concept for young people to understand. The emphasis in the book is on the use of natural resources rather than strictly preservation of resources.

Agricultural and science concepts are carefully articulated. Relationships and applications are shown. Hands-on activities are suggested that allow students to experience the science of agriculture. Personal development is also included, because the development of personal skills is important to the success of people in agricultural occupations.

The Earth and AgriScience is a full-color book designed for use in the middle school grades. The grades in which it is used may range from grade 5 to grade 8 or, in some cases, grade 9. Computer readability analyses have been made in an attempt to keep the reading level appropriate. Concepts are carefully introduced with the needs and interests of middle school students in mind. Numerous illustrations are used to enrich the content of the book.

The publisher and authors have attempted to produce a student text that is both student and teacher friendly. You will find *The Earth and AgriScience* a useful new approach in the education of students.

PURPOSE OF THE TEACHER'S MANUAL

The Teacher's Manual to accompany *The Earth and AgriScience* is designed to help the teacher in the instructional process. A systematic approach is used to merge education in the environment, science, and agriculture. An overview of the contents of the book is presented in the Teacher's Manual. Suggested teaching strategies include using motivational approaches, as well as applying the content through activities included in both the text and this manual.

Many different resources can be used in providing instruction. *The Earth and AgriScience* student text is the main component in the instructional program. Suggestions for incorporating classroom and laboratory activities are included. Some of these will extend beyond the school campus and into the local community.

FORMAT OF THE TEACHER'S MANUAL

The format of the Teacher's Manual is easy to follow. The manual is divided into various sections that will help the teacher in gaining student interest, providing instruction, reviewing and reteaching, evaluating, and incorporating hands-on activities in this combined environment- and science-based approach to introductory agriculture.

The Teacher's Manual is divided into 27 chapters corresponding to the chapters in *The Earth and AgriScience*. Each chapter is further divided into several sections as follows:

Chapter Summary. This section contains a short summary of the chapter in the text. Additional information may also be included to provide background content to help the teacher in the instructional process.

Instructional Objectives. The objectives for each chapter in the text are stated in measurable and observable terms. These are the behaviors that learners should have upon completion of the instruction.

Interest Approach, Instructional Strategies, and Teaching Plans. A possible interest approach to get the attention of students and motivate them to learn is included. Strategies for planning and teaching the content are suggested. The focus is always on helping the students achieve the objectives. Strategies for higher-order thinking skills are often included. Procedures for the articulation and integration of instructional content with other areas, such as science and history, are included as appropriate.

Review and Evaluation. Suggestions for reviewing the content and reteaching as needed are included. The focus is to assure that the students have mastered the objectives. Suggestions for evaluation are included.

Additional Resources. This section contains information about other instructional resources that the teacher may find useful.

Answers to Questions and Chapter Self-Checks. The Teacher's Manual includes answers to the end-of-chapter "Questions" and the "Chapter Self-Check" evaluation.

OVERVIEW OF THE EARTH AND AGRISCIENCE

The Earth and AgriScience presents fundamental principles and concepts associated with providing the food, clothing, and shelter that people need. It explains the broad nature of renewable and non-renewable natural resources and how these can be used to meet the needs of people. Science principles are included to emphasize the role of science in agriculture. Areas of the environment are included to stress the importance of taking care of the earth. Career success begins with selecting the right area of work and developing the needed personal skills.

Emphasis is on practical learning and activities. The content is designed to encourage students to take additional courses in agriculture the following years. Students are also encouraged to carry what they learn beyond the classroom and into their homes and communities.

The Earth and AgriScience is divided into five parts and 27 chapters. These are organized as follows:

Part One: The Earth's Resources

Part One introduces students to the renewable and non-renewable natural resources found on the earth and how these meet the needs of people. Ways of using the resources are explained. The emphasis is on sustaining the natural resources for long-term use and benefit. This part has three chapters:

    Chapter 1 — What Our Earth Provides
    Chapter 2 — How We Use the Earth's Resources
    Chapter 3 — Sustaining the Earth's Resources

Part Two: Our Food and Fiber

The five chapters in Part Two present the needs of people and how agriculture helps meet the needs. Emphasis is on proper use of resources and science-based information in providing food, clothing, and shelter. The chapters are:

Chapter 4 — What We Need
Chapter 5 — Meeting Our Needs
Chapter 6 — Beyond Food: Ornamental Horticulture
Chapter 7 — Using Forest Resources
Chapter 8 — Getting Products to Us

Part Three: Using AgriScience

This part presents the fundamentals of science in agriculture. How these fundamentals relate to our environment are included. The application of science through technology is introduced. Specific information is included on plants, animals, and power technology. The seven chapters in Part Three are:

Chapter 9 — Using Science in Agriculture
Chapter 10 — Sustaining Ecosystems
Chapter 11 — Growing Plants
Chapter 12 — Raising Animals
Chapter 13 — Keeping Animals as Pets
Chapter 14 — Using Power Machinery
Chapter 15 — Helping Plants and Animals Grow

Part Four: Protecting the Earth

Part Four has seven chapters that cover environmental conservation. The importance of maintaining a good environment and conserving the natural resources is stressed. Content on preventing pollution and recycling is presented.

Chapter 16 — Keeping a Good Environment
Chapter 17 — Keeping Our Soil
Chapter 18 — Keeping Our Water
Chapter 19 — Keeping Our Air
Chapter 20 — Keeping Our Wildlife
Chapter 21 — Preventing Pollution
Chapter 22 — Recycling

Part Five: Finding What I Want

This part includes important information to help young people set and attain career and success goals. Job expectations and education needs are covered. The role of personal skills, civic responsibility, and maintaining a healthy body are included. All of these areas are intended to help people live a long, productive life. The five chapters are:

Chapter 23 — Careers in AgriScience
Chapter 24 — Getting the Education
Chapter 25 — Developing Personal Skills
Chapter 26 — Being a Good Citizen
Chapter 27 — Keeping Healthy

CHAPTER ORGANIZATION

All chapters in *The Earth and AgriScience* follow a similar format. The arrangement of the chapter material is important for systematic instruction.

Each chapter begins with a short introduction that provides motivational background information on the content. This is followed by a list of several questions (usually three to five questions) that serve as objectives for the chapter. "AgriScience Vocabulary" words are listed following the "Objectives."

Systematic Instruction

The chapter content is sequentially organized into headings based on the objectives and essential content. The content has been carefully organized to make the major content and organization prominent to the learner. Color photographs and other illustrations have been used to make the content highly salient for students.

Each chapter concludes with a section entitled "Reviewing", which is divided into "Main Ideas" and "Questions." Evaluation involves using the "Chapter Self-Check." The review questions can also be used as tools for evaluation and reteaching. The "Exploring" section includes additional activities that reinforce learning and provide hands-on application of the content.

Career Information

Each chapter contains a "Career Profile," with one chapter having two. This involves a short description and a photograph of a person in an occupation that is related to the content of the chapter. The descriptions contain information about the nature of the work in the occupation, as well as the educational requirements for the occupation. The book contains 28 career profiles.

USING THE EARTH AND AGRISCIENCE IN THE INSTRUCTIONAL PROGRAM

The Earth and AgriScience is designed for maximum flexibility in agricultural education. It is intended as the book that students would have in their first agriculture or agriscience instructional program. The book can be used in classes that are prerequisites for agriscience, agriculture, or environmental science in succeeding years. It has been designed for flexibility in the middle school program.

The Earth and AgriScience is organized for adapting to a variety of curriculum patterns. Classes that are offered as modules of six or nine weeks can use one or two parts, while those lasting a semester can use additional parts depending on preferences and the time available. The book is adaptable to curriculum patterns that provide modules in agriscience each year in the middle school curriculum. For example, Part One could be used for a short module in grade 6, Parts Two and Three could be used for nine weeks or longer in grade 7, and Parts Four and Five for a semester in grade 8.

The book has been designed as a student- and teacher-friendly introduction to agriscience.

Chapter 1

WHAT OUR EARTH PROVIDES

CHAPTER SUMMARY

The earth is the source of all the food, clothing, and shelter that people need to live long, productive, and happy lives. Resources are the naturally occurring things that people use from their environment. Environment is all of the factors that affect our lives—the things around us.

Conservation is using resources so that we get the most out of them. Some resources are renewable; others aren't. Renewable resources can be replaced, but we must work to see that replacement occurs. Non-renewable resources are gone when they are used.

Agriscience is the use of science in studying and working in agriculture. It includes all areas of science, with particular emphasis in earth science—the study of the earth and its atmosphere. Yes, the air immediately above the earth is a part of the earth!

INSTRUCTIONAL OBJECTIVES

The objectives for this chapter are intended to help learners understand the importance of natural resources. The objectives are stated as questions in the student text.

Upon completion of chapter 1, the student will be able to:

1. Describe the earth's resources.
2. Distinguish between renewable and non-renewable natural resources.
3. Explain how earth science and agriscience helps people use natural resources more efficiently.

INTEREST APPROACH, INSTRUCTIONAL STRATEGIES, AND TEACHING PLANS

Many teaching procedures can be used with this chapter. Several suggestions are included here.

Begin the interest approach by having the students read the introductory section of the chapter. Ask them to make a list of the five most important things that they get from the earth. Ask them to explain how these relate to the environment. (In some cases, the items the students list may cause pollution. In other cases, they may be made of resources that are depleted.)

Following the interest approach, review the "Objectives" for the chapter. (These are stated as questions on page 4 in the student text.) After the questions to be answered have been identified, the teacher may refer the students to the list of "AgriScience Vocabulary" words. These words are the most important terms covered in the chapter.

Students are now ready to move into the content of the chapter. Strategies here vary with the needs and interests of the students. Learning will be more permanent if the students read the material before it is discussed and otherwise presented in the class. Some teachers will use presentation methods to cover chapter content. Other teachers will have students explain different sections of the chapter. Stressing the importance of the earth's resources to the well being of people is a major goal of this chapter. Students may be asked to mention examples studied in other classes that relate to the chapter. Teachers may want to articulate the instruction in the chapter with that of teachers in science or other areas.

REVIEW AND EVALUATION

Review can be achieved by having students use the "Reviewing" section of the chapter. The students should read the "Main Ideas." Calling on individual students to summarize the "Main Ideas" section will be useful in helping learners internalize their studies.

The "Questions" can be used for review or for evaluation. Some teachers have students write the answers during supervised study in the class. Other teachers will assign the questions as homework. Orally discussing the questions will help in review, evaluation, and reteaching.

Evaluation can include completion of the "Chapter Self-Check", written or oral tests, and observation of performance on activities.

ADDITIONAL RESOURCES

Information about natural resources that are found in the local area will be useful. Particularly important are oil wells, coal mines, ore mines, and other non-renewable natural resources. The local area will have varying renewable natural resources, and these should be included as appropriate. This information is usually available from a university or a state resource conservation agency.

ANSWERS

QUESTIONS

1. What are the earth's resources?

Resources are the things that people use to live. Some are essential to human life, such as the air we breathe. Others, such as oil, are used to make life more comfortable.

2. How is the environment important to plants and animals?

The environment is important to plants and animals because they get what they need to live and grow from it.

3. Why is conservation important? How is it different from preservation?

Conservation is using resources so that we get the most out of them. Preservation involves maintaining resources and not using them.

4. What is the major difference between renewable and non-renewable natural resources? List three examples of each.

Renewable natural resources can be replaced when they are used. Non-renewable natural resources cannot be replaced when they are used. Examples of renewable natural resources are: water, soil, air, wildlife, and forests. Examples of non-renewable natural resources are oil, coal, iron, and copper.

5. What can people do to make non-renewable natural resources last longer?

People need to know about earth science and agriscience. They need to know how to prevent loss of resources and damage to them. Knowing conservation practices is important.

6. Why is the structure of the earth sometimes compared to an apple? Explain.

The earth has a core and a surface similar to an apple. The earth's surface is represented by the apple's skin. The white part of an apple represents various layers within the earth. The area surrounding the earth and apple is the atmosphere.

7. Why should people know about earth science?

Knowledge of earth science helps people conserve the use of natural resources. It helps keep a good environment.

8. Why is agriscience important to people?

Agriscience helps people use the earth's resources to meet their needs. People are able to increase productivity from the resources.

CHAPTER SELF-CHECK

1=D, 2=C, 3=G, 4=H, 5=E, 6=B, 7=A, and 8=F.

Chapter 2

HOW WE USE THE EARTH'S RESOURCES

CHAPTER SUMMARY

Natural resources are used to meet the needs of people. Life has changed from hunting and fishing to a modern life style that requires the intensive use of natural resources to meet the needs of the earth's ever-growing population.

The renewable natural resources include water, soil (not dirt!), air, forests, and wildlife. All of these are needed for human life and comfort. Through conservation practices, these resources can be renewed. In some cases, a lot of time and effort may be required. It is better to avoid damaging the resources and unnecessarily using them than it is to try to replace them.

The non-renewable natural resources include fossil fuels and minerals. Fossil fuels include coal, oil, and natural gas. They are known as fossil fuels because they were once living plants and animals. Minerals are important for plants and animals to grow, as well as to use in manufacturing. Iron ore is an important mineral that is used to make steel and other important construction materials. And many others could be named!

INSTRUCTIONAL OBJECTIVES

The objectives for this chapter are intended to help learners understand the importance of natural resources. The objectives are stated as questions in the student text.

Upon completion of chapter 2, the student will be able to:

1. Explain why proper use of the earth's resources is important.
2. Describe how renewable natural resources are used.
3. Describe how non-renewable natural resources are used.

INTEREST APPROACH, INSTRUCTIONAL STRATEGIES, AND TEACHING PLANS

A variety of teaching strategies can be used to teach the responsible use of the earth's natural resources. Several suggestions are included here.

Begin the interest approach by having the students read the introductory section of the chapter. Ask students to explain why when something is used it is taken away from the earth. Also have them explain why using resources changes the earth. Following the interest approach, present the "Objectives" for the chapter. You may wish to have the students review the questions that they should be able to answer. These are parallel with the objectives listed above.

After the objectives have been presented, begin the content of the chapter. Have the students read sections of the chapter in class or as homework. Discuss each section after it has been read. Begin discussion by having the students tell how life has changed from centuries ago. Ask the students how animal and plant domestication changed peoples' lives. Stress the new developments. Move into a discussion of the different natural resources. Begin with renewable natural resources and then move to the non-renewable natural resources. Cover the content one resource at a time. Use questions that relate the resource to the lives of the students. With soil, particularly emphasize that soil isn't dirt. Write the key terms in each section on the chalkboard. Follow the terms with brief definitions that the students give during the discussion. Have the students write the terms and definitions in a notebook.

REVIEW AND EVALUATION

Upon completion of the content of the chapter, have students read the "Main Ideas" part of the "Reviewing" section. Call on different students to summarize the main ideas. After the main ideas have been reviewed, have students answer the questions either as homework or as supervised study in the classroom.

Evaluation can involve using the "Chapter Self-Check", tests that are made by the teacher, how students respond in class, and how students participate in the activities in the "Exploring" section.

Student performance on the review and evaluation may indicate the need for reteaching. If so, use the questions at the end of the chapter or have students use the term list at the beginning of the chapter and write down the definitions of the terms.

ADDITIONAL RESOURCES

Additional resources may be selected to give information about the natural resources in the local community. Contact the local Cooperative Extension Service office, Soil Conservation Service, or other agencies.

ANSWERS

QUESTIONS

1. Why are natural resources important to people?

Natural resources are important to people because they are used in our daily living. The natural resources are used to provide food, clothing, and shelter.

2. Why did people start domesticating plants and animals?

People found it easier to have food when they wanted it if they grew it. Hunting wasn't required.

3. What is water? Why is it important in agriscience?

Water is a liquid made of two elements: hydrogen and oxygen. The formula for water is H_2O. Water is important because it is essential to life. Both plants and animals require water to live.

4. What is the difference between dirt and soil?

Soil is the outer layer of the earth's surface. It is made of various materials that provide nutrients for plants to grow. Dirt is something that is unclean. Dirt can be soil if the soil is where you don't want it, such as on your hands.

5. How is soil formed?

Soil formation is a continuous process. It is formed from parent material, including rock and decaying plants and animals. Decaying plants and animals form organic matter.

6. Why is air important to plants and animals?

Air contains oxygen and carbon dioxide. Animals and plants use oxygen to live. Plants use carbon dioxide and release oxygen.

7. Why are forests important?

Forests are important because they contribute many resources, such as wood and wildlife. Forests improve air and soil quality. They help prevent erosion and provide food for all kinds of wildlife. Forests are also used for recreation.

8. **What are fossil fuels? Name three examples.**

Fossil fuels come from plants and animals that were alive millions of years ago. Examples are: coal, natural gas, and oil.

9. **How are minerals important to plants and animals?**

Small amounts of minerals are needed for plants and animals to grow and be healthy. When they don't get enough minerals, they get sick.

CHAPTER SELF-CHECK

1=J, 2=D, 3=E, 4=B, 5=A, 6=C, 7=G, 8=F, 9=I, and 10=H.

Chapter 3

SUSTAINING THE EARTH'S RESOURCES

CHAPTER SUMMARY

This chapter focuses on careful use of the earth's resources so that people have them for many years to come. Sustainable resource use means that resources are used in such a way they last a long time. Agriscientists have developed a special term: sustainable agriculture, which is producing plants and animals so that the ability to grow them isn't lost. Sustainable resource use provides for a good quality of life for people.

People sometimes use the earth's resources unwisely. This causes damage. Wastes of all kinds are created every day. The major ones are solid wastes, waste water, and gases and fumes. Some of these pollute the environment.

Pollution and damaging the earth in other ways has a big impact on the earth. The earth is beautiful, and this is destroyed with pollution. Soil can be lost to erosion. Deforestation results in lower air quality. Acid rain damages many things. Global warming, smog, and contaminated water lower the quality of life on the earth. Many kinds of wastes can contaminate water and endanger wildlife and destroy wetlands.

INSTRUCTIONAL OBJECTIVES

The objectives for this chapter are intended to help learners understand the importance of sustaining the earth. The objectives are stated as questions in the student text.

Upon completion of chapter 3, the student will be able to:

1. Describe sustainable resource use and sustainable agriculture.

2. Explain how quality of life is related to resource use.

3. Identify sources of damage to the earth's resources.

4. Explain how damage affects the earth's resources.

INTEREST APPROACH, INSTRUCTIONAL STRATEGIES, AND TEACHING PLANS

A wide range of approaches can be used in teaching this chapter. The approaches should be selected on the basis of student needs and interests. Resources in the local community, particularly those with conservation problems, may be important in motivating students.

The interest approach can begin by having students read the introductory section of the chapter. Ask students if electricity has gone off in their homes recently. What happened? Ask the students to explain how using up all resources can be compared to the electricity going off.

Following discussion of the introductory section, present the objectives for the chapter. This can involve reviewing the questions in the text that cover the chapter content.

More thorough learning will result when students are actively involved in reading and responding about the content of the chapter. Have them read sections at a time. Follow their reading with discussions. Particularly ask student questions that help them relate the chapter content to life in their local community. Emphasis should be on sustainability. Spend sufficient time in presenting, discussing, and reteaching this concept to be sure that all students have grasped its meaning. Sustaining the earth's resources is related to the quality of the lives that people will have in the future. Some of the ways people produce wastes may be surprising to the students, such as taking a bath produces waste water. Discuss the ways damage affects the earth. Ask students if they know of examples of the damage.

REVIEW AND EVALUATION

The "Reviewing" section at the end of the chapter can be used for both review and evaluation. Have students read the "Main Ideas" section and call on several students to summarize what they have read. Assign the questions as homework or use them for supervised study. In some cases, the questions may be discussed in class.

Evaluation can involve using the review questions, as well as the "Chapter Self-Check." Teacher-made tests and the computer test bank can also be used in evaluation. Reteach any content that is needed on the basis of the evaluation.

ADDITIONAL RESOURCES

Specific examples where sustainability has been practiced in the local community should be identified. The local Soil Conservation Service office or the Cooperative Extension Service office may be able to assist in identifying the farms that are using sustainable agriculture practices. If possible, make a field trip to one of the farms to observe what is done. Have the farm manager explain sustainable agriculture.

ANSWERS

QUESTIONS

1. **Why is sustainable resource use important?**

   Sustainable resource use is important because it eliminates waste and makes the resources available for a long time in the future.

2. **What is the role of the agriscientist in sustainable agriculture?**

   Agriscientists study ways to make agriculture sustainable. Without their work, people wouldn't know what to do. New ways of growing crops and raising animals without chemicals are developed through research.

3. **Why is conservation important in quality of life?**

   Quality of life means that people have a healthy environment in which to live. They enjoy life a long time. The needs of people are met.

4. **What kinds of waste damage resources? How do they cause damage?**

   The kinds of waste are: solid waste; waste water; and gases, fumes, and air. These cause damage by polluting our environment. Wastes build up and cause problems. Some of the wastes are toxic and kill or injure living things.

5. **What are the sources of pollution?**

   Daily living activities create pollution. Homes, factories, and farms produce pollution. Sources of pollution can be regulated or eliminated.

6. **How is pollution prevented?**

   Pollution can be prevented by properly disposing of wastes. Trash can be recycled. Factories can control the release of smoke. Farms can use practices that don't pollute.

7. **How does pollution change the earth?**

   Pollution destroys the beauty of the earth. It results in the loss of soil. Deforestation reduces the amount of oxygen produced for the air. Acid rain, smog, global warming, and contaminated water may injure living things, as well as machinery and equipment. Some species of plants and animals are endangered. Special areas known as wetlands may be destroyed.

8. **What can be done to keep the earth beautiful?**

   Don't pollute. Practice sustainable resource use. Properly dispose of wastes.

CHAPTER SELF-CHECK

1=B, 2=H, 3=F, 4=G, 5=A, 6=D, 7=C, and 8=E

Chapter 4

WHAT WE NEED

CHAPTER SUMMARY

People have basic needs: food, clothing, and shelter. Food is the solid and liquid materials that people eat and drink. Food provides nutrients. Nutrients are substances that provide nourishment. People need six nutrients: carbohydrates, proteins, fats, vitamins, minerals, and water. The food we eat must provide these nutrients.

The food guide pyramid provides important information to help people eat what they should. Many processed food products have labels that describe the kind of product, weight, date, and name of the company that made it as well as the nutrients that are provided. People want food that is nutritious and safe to eat.

Clothing needs include garments, ornaments, and accessories. Garments are the coverings that people wear to protect the body and improve appearance. Ornaments are worn to beautify the body. Accessories are worn to supplement the basic clothing and include necklaces and rings. Clothing choices vary with cultures and income. Clothing also tells something about the status of a person. Successful business people are expected to dress a certain way. Most ready-made clothing is made from cotton, wool, linen, and synthetic materials.

Shelter is the housing that people have. The nature of housing ranges from tiny apartments in large buildings to single family homes and mansions. Shelter is made of resources from the environment. Wood, metal, clay, glass, and other materials are commonly used.

INSTRUCTIONAL OBJECTIVES

The focus of chapter 4 is on helping young people understand how their basic needs are met. This involves using resources that are found on the earth.

Upon completion of chapter 4, the student will be able to:

1. List the basic needs of humans.
2. Explain the important food needs of humans.
3. Describe how humans want their food.
4. Explain the clothing needs of humans.
5. Explain the shelter need of humans.

INTEREST APPROACH, INSTRUCTIONAL STRATEGIES, AND TEACHING PLANS

Since this chapter relates closely to the daily lives of the students, the instructional strategies should relate to their needs, preferences, and experiences.

The interest approach can begin by having them read the introductory section of the chapter. Ask the students to tell about some of the things they enjoy most in life. Are these related to food, clothing, and shelter? Ask the students if they know of someone who doesn't have sufficient food, clothing, or shelter, such as the homeless or indigent.

After the interest approach, review the "Objectives" for the chapter. This may include reviewing the five questions in the text that are answered in the chapter. These questions are parallel with the objectives listed above.

Involve students in reading and synthesizing the content of the chapter. Have them read a section at a time and discuss what they have learned. Write key terms and concepts on the chalkboard. A good activity is to have them compare what they ate the previous day with the food guide pyramid. How does their nutrition correlate with the recommendations? Bring samples of food labels for the students to review and interpret. In some cases, they may need help in understanding the label. Assess several different labels to determine which of the food products contains more of the nutrients people need for good health.

The section on clothing will likely be of high interest to students because of the styles in which they are interested. Spend some time looking at clothing labels to determine the product used to make the garments, as well as instructions to follow in caring for them.

Housing is of high interest to some people. In a few cases, students may be from housing projects and sensitive to where they live. Every effort should be made to in no way embarrass any student because of the kind of housing that they have. It isn't a fault of their own if their housing is not up to standard. Have students read and discuss the sections of the chapter on housing, just as with other sections.

REVIEW AND EVALUATION

Use the chapter "Main Ideas" section and "Questions" in reviewing. The process of review may reveal the need for reteaching and provide evaluative information. Activities at the end of the chapter in the section entitled "Exploring" will be useful.

Evaluation can be with the "Chapter Self-Check", "Questions", teacher-made tests, observation of overall class performance, computer-generated tests, and other means.

ADDITIONAL RESOURCES

Additional resources for this chapter include brochures on the food guide pyramid, clothing selection and care, and home maintenance. These are available from the Cooperative Extension Service office in your local area. The school food service may also have useful materials on food. In some cases, the school lunchroom manager would be a good resource person by explaining how cafeteria food is prepared to meet nutritional needs.

ANSWERS

QUESTIONS

1. **What nutrients do people get from food? Name two foods that are good sources of each nutrient?**

    Food provides carbohydrates, proteins, fats, vitamins, and minerals. There are many examples of each. Refer to the food guide pyramid for the examples.

2. **What is a label? What information does a label on food give?**

    A label is a small piece of paper, cloth, or metal with printed words that tell about a product. Food labels give the following information: kind of product, weight, serving size, name of the company that made it, and information about the nutritional content of the food.

3. **How is the food guide pyramid useful in eating right?**

    The food guide pyramid shows the proper amount of each food group to eat each day. People who follow the food guide pyramid get the nutrients they need.

4. **What are garments and accessories?**

    Garments are coverings that people wear on their bodies. Accessories are the things that they wear to supplement the basic clothing or garments.

5. **What materials are used to make clothing?**

    Clothing is made from natural and synthetic materials. The natural materials include cotton and wool. The synthetic material includes plastic and glass.

6. **What kinds of shelter do people use?**

    Houses, apartments, mobile homes, and condominiums are common kinds of shelter.

CHAPTER SELF-CHECK

1=H, 2=G, 3=E, 4=F, 5=B, 6=A, 7=D, and 8=C.

Chapter 5

MEETING OUR NEEDS

CHAPTER SUMMARY

People no longer hunt and fish to meet most of their needs. Agriculture is a big industry that provides food, clothing, and shelter to meet the needs of people.

Agriculture is producing and using plants and animals to meet the needs of people. Farming is the actual production of plants and animals on the land. The place where all of this takes place is often known as a farm. What happens on a farm has changed tremendously. Agriculture is commercial, meaning that farmers produce plants and animals to sell to others. Many people are involved in agricultural occupations in off-farm places to help get the produce of farms to consumers in the forms that they want.

A huge agricultural industry has developed. This includes all of the steps in getting food, clothing, and shelter to people in the forms they want and when they want them. The supplies and services sector of the agricultural industry provides the inputs that farmers need and use, such as fertilizer and seed. The marketing sector includes all of the activities in getting a product to people who will use it. The term, agribusiness, is often used to describe the combined supplies and services and marketing sectors of the agricultural industry.

INSTRUCTIONAL OBJECTIVES

The objectives of this chapter are intended to help students understand the broad nature of agriculture, including the agribusiness sectors. The objectives are stated as questions in the student text.

Upon completion of chapter 5, the student will be able to:

1. Define agriculture and farming.
2. Explain the major sources of food, clothing, and shelter.
3. Describe the agricultural industry and agribusiness.
4. List and explain three major areas of the agricultural industry.

INTEREST APPROACH, INSTRUCTIONAL STRATEGIES, AND TEACHING PLANS

A wide range of teaching strategies can be used with this chapter. Several suggestions are included here, with emphasis on using the textbook materials to help students master the objectives.

For the interest approach, have students read the introductory section of the chapter. Ask if any of them have been hunting or fishing and how they would like to depend on hunting and fishing for their food, clothing, and shelter. In most cases, the students would not likely have enough of these to meet their basic needs. Explain that this is the reason we have agriculture: Agriculture helps to more adequately meet the needs of people than hunting and fishing would.

Following the interest approach, review the objectives with the students. Some teachers may also wish to review the list of "Agriscience Vocabulary" terms at this time. Other teachers may have the students use the list and develop written definitions during a time of supervised study.

Various sections of the chapter can be read in class or as homework. The content can be discussed and outlined on the chalkboard. Emphasis should be on the key meanings of farming, agricultural industry, and agribusiness. The supplies and services and marketing sectors should be carefully explained. Students can provide numerous examples of each that they see in their local communities.

REVIEW AND EVALUATION

Use the "Reviewing" section of the chapter to aid in review. Have students read the summary and orally discuss the key information. Students can answer the questions at the end of the chapter in writing or orally in class. The "Agriscience Vocabulary" terms can also be used in the review process.

Evaluation may involve using the "Chapter Self-Check", "Reviewing Questions", or teacher-made tests. A computer test bank is available from Interstate Publishers, Inc., for the book.

Reteaching may involve additional discussion of the important concepts in class or the use of the "Agriscience Vocabulary" list, the "Reviewing Questions", or the "Chapter Self-Check".

One or more of the "Exploring" activities may be used in review and evaluation.

ADDITIONAL RESOURCES

The "Exploring" activities at the end of the chapter provide excellent suggestions to help expand the content of the chapter. These activities also reinforce chapter concepts and review the overall content of the chapter.

Statistical information about agriculture in the local area may be useful with this chapter. Students will benefit from knowing about the agricultural industry in the community in which they live. The information may be found in the agriculture census summary or available from the state commissioner of agriculture, economic development office, or local Chamber of Commerce.

ANSWERS

QUESTIONS

1. **What is agriculture? How is farming a part of agriculture?**

   Agriculture is producing and using plants and animals to meet the needs of people. Farming is the use of land to produce the plants and animals. Farming is a part of agriculture, and agriculture is greater than farming.

2. **Why is agriculture important?**

   Agriculture is important because it is how most of the food, clothing, and shelter are produced. It also provides jobs for a lot of people.

3. **What are the sources of food, clothing, and shelter?**

   Food and clothing are from both plant and animal sources. Most of the material to build a house is from plants and minerals, such as clay for brick.

4. **What is agricultural industry? How is it different from agriculture?**

   Agricultural industry is all of the activities in providing food, clothing, and shelter. Agriculture is the production of the plants and animals that are made into useful products.

5. **Describe the three major areas of the agricultural industry.**

   The three major areas of the agricultural industry are: agricultural supplies and services, production agriculture, and agricultural marketing. Supplies and services include the inputs used on the farms, such as feed, seed, and fertilizer. Production agriculture is the same as farming. Agricultural marketing is all of the steps that get products to people the way they want them. It includes processing farm produce.

6. **What major changes have taken place in agriculture?**

   Modern agriculture involves tractors, computers, and lasers. This is a lot different from the past where many people worked in fields using hand tools. A lot of the agriculture work has moved from the farm to factories and other places that get farm products ready for people to use.

CHAPTER SELF-CHECK

1=H, 2=C, 3=D, 4=B, 5=A, 6=E, 7=F, and 8=G.

Chapter 6

BEYOND FOOD: ORNAMENTAL HORTICULTURE

CHAPTER SUMMARY

Ornamental horticulture is producing plants for their beauty. Aesthetics means that the plants appeal to the senses of people. Ornamental horticulture is a part of agriculture but it is different from producing crops and livestock.

Ornamental horticulture has two major areas: floriculture and floral design and landscaping. Floriculture and floral design deal with flowers

—19—

and foliage. Flowers are the attractive blossoms on plants. Foliage is the attractive leaves and stems of plants. Landscaping is using plants to have a more attractive outside environment. Landscaping involves using many different kinds of plants and growing them in carefully planned designs. Lawns, shrubs, and flowering plants are often used in landscapes. Interior plantscaping is used with the inside areas of malls, office buildings, and other places.

A lot of science is used in horticulture. Plants require water, light, soil or medium, and nutrients. Greenhouses, shade houses, and growing beds may be used to get the plants produced in a timely manner.

Horticulture provides many benefits. Several are: enjoyment, as a hobby, to express emotion, and for therapeutic purposes. Many people have rewarding careers in horticulture.

INSTRUCTIONAL OBJECTIVES

The objectives for this chapter are intended to help students understand the nature of horticulture and the different areas that are involved. The objectives are stated as questions in the student text.

Upon completion of chapter 6, the student will be able to:

1. Describe ornamental horticulture.
2. Explain how ornamental horticulture relates to agriscience.
3. List and explain the benefits of ornamental horticulture.

INTEREST APPROACH, INSTRUCTIONAL STRATEGIES, AND TEACHING PLANS

Teachers are creative. They are able to adapt content and objectives to the needs of students and other situations that exist in the classroom. Several suggestions are presented here.

Begin the interest approach by having students read the introductory part of the chapter. Ask them if they send flowers on special days. Have them name the days. Ask them if they like to get flowers. Lead the students to understand that both boys and girls like flowers and areas of beauty in parks, on golf courses, lawns, and other places.

The plan for covering the content of the chapter should focus on mastering the objectives and enhancing the reading skills of students. This will include having the students read the chapter during class or as homework. The various sections of the chapter should be discussed in class, with key terms and definitions written on the chalkboard.

The school grounds or a nearby home or business that has a lawn and landscape can be used for a short walking field trip. Identify the common plants used in the lawn and as shrubs and flowering plants. Explain the importance of having the right plant in the right place. Ask, "What happens if a plant that grows tall is planted in front of a window?" Naturally, the plant will obstruct the view from the window. This chapter may be used as a good opportunity to start a few small plants in the classroom as a part of interior plantscaping the room.

REVIEW AND EVALUATION

Use the "Reviewing" section at the end of the chapter to help with review. The students can read and orally explain the content of the "Main Ideas." The "Questions" can be answered in writing as homework or orally during class.

The "Chapter Self-Check" can be used in evaluation, as well as teacher-made tests, tests from the computer test bank, and general class performance of the students.

Reteaching may be needed with some of the concepts, based on the evaluation. Some teachers use the list of terms in reteaching. Students may be required to write or orally explain the meaning of each "Agriscience Vocabulary" term.

ADDITIONAL RESOURCES

Helpful additional resources include local information about horticulture, such as brochures on lawns, shrubs, and flowering plants. These are available from the local office of the Cooperative Extension Service or a garden club. Nurseries and garden centers may have useful information.

The "Exploring" section at the end of the chapter will help direct students toward additional resources.

ANSWERS

QUESTIONS

1. **What is aesthetics? Name a place in your community that has good aesthetics and tell why you chose this place.**

   Aesthetics means that something appeals to the senses of people. (The place that students select may be a park, lawn, or other location based on the local community.)

2. **What is ornamental horticulture?**

   Ornamental horticulture is producing plants for their beauty.

3. **Why are flowers and foliage used to make bouquets?**

   Flowers and foliage are used to make bouquets because they are attractive. People like the beauty of flowers.

4. **Why are turf, shrubs, and flowering plants used in landscaping?**

   Each appeals to people and serves a special purpose in a landscape.

5. **What is interior plantscaping? Where is it used?**

   Interior plantscaping is growing plants in garden-like areas inside of buildings. It is used in malls, office buildings, hotel lobbies, government buildings, and many other places.

6. **What structures are used in ornamental horticulture?**

   The structures used in ornamental horticulture are greenhouses, growing beds, and shade houses.

7. **What do plants need in order to grow?**

   Plants need temperature, light, air, water, and growing medium in order to grow. The growing medium must contain needed nutrients.

8. **What are the benefits of ornamental horticulture?**

   The benefits of ornamental horticulture are: enjoyment, hobby, express emotion, therapeutic value, and career opportunities.

CHAPTER SELF-CHECK

1=B, 2=C, 3=D, 4=G, 5=E, 6=F, 7=A, and 8=H.

Chapter 7

USING FOREST RESOURCES

CHAPTER SUMMARY

Forestry is managing forests so that good quality products are produced. Forests are more than trees! Forest resources include all of the living things that grow in forests — insects, birds, squirrels, vines, mushrooms, and many plants and animals in addition to the trees. Forests are important because they produce valuable products, provide recreation, and help maintain a good environment.

Tree farms are taking the place of forests as sources of wood products. More than 20,000 different kinds of trees are found in the world, with about 1,000 found in North America. Naming and classifying trees is known as dendrology. Only a few species of trees are produced on tree farms. These include species of pine, fir, spruce, walnut, maple, and oak.

Trees have three major parts: crown, trunk, and root. Products are made from all parts of a tree. Of course, the trunk is most valuable as a source of lumber, paper, plywood, and other products.

Operating a tree farm is much like a crop farm. All of the cultural conditions to help the trees grow must be provided. A major difference between trees and most crops is that trees take several years to grow to a size where they can produce useful products. The media have given a lot of coverage to the importance of rain forests with the notion that the forests should not be cut. This chapter addresses forestry as the commercial farming of trees.

INSTRUCTIONAL OBJECTIVES

The objectives of this chapter are designed to help students understand the nature of forestry. Emphasis is on growing trees on tree farms.

Upon completion of chapter 7, the student will be able to:

1. Explain forest resources and why they are important.
2. Draw and label the parts of a tree and list one product from each part.
3. Explain how trees are farmed.
4. List the major products from trees.

INTEREST APPROACH, INSTRUCTIONAL STRATEGIES, AND TEACHING PLANS

The instructional strategies for chapter 7 will vary with the forestry found in the local area. Some communities will have considerable emphasis in forestry; others will have very little. Regardless, the products of forests are important everywhere.

Begin with the interest approach by having students read the introductory section of the chapter. Have them prepare the list as described in the second paragraph of the chapter. Discuss the lists in class. Develop a combined list of all class members on the chalkboard.

Move from the interest approach into the objectives. These can be reviewed using the questions in the chapter or written on the chalkboard for the students.

Following the objectives, begin the content of the chapter. Have students read the chapter as homework or as supervised study. Discuss each section of the chapter and write key terms and definitions on the chalkboard.

REVIEW AND EVALUATION

Review the content of the chapter using the "Reviewing" section. Begin by having students read the section, "Main Ideas." Call on various students in the class to orally summarize what they have read.

Use the "Questions" for oral review in the class or have students write the answers during supervised study or for homework.

Evaluation can involve using the "Chapter Self-Check", as well as teacher-made tests.

Reteaching can be based on observations of the students during discussion of the "Main Ideas" and "Questions." The "Agriscience Vocabulary" list may also be used in reteaching.

ADDITIONAL RESOURCES

Information about forestry in the local area will be helpful. Some states have local foresters who can provide materials on tree species, fire prevention, harvesting, and other areas. Timber and forestry industries in the local community may also have materials that would be useful.

ANSWERS

QUESTIONS

1. What is the difference between a forest and a tree farm?

Forests and tree farms aren't the same. Forests are large areas of land covered with trees. Tree farms are farms where trees are grown. Tree farms use specific kinds of trees and care for them in certain ways.

2. **What are forest resources?**

   Forest resources are all of the living things that grow in forests.

3. **What are three important things we get from forests? Briefly explain each.**

   Three important things we get from forests are products, recreation, and maintain the environment. Products include lumber, plywood, and paper. Recreation includes camping, hunting, hiking, and picnicking. Forests release oxygen into the air to help maintain the environment.

4. **What does a tree farmer consider in selecting which trees to grow?**

   Tree farmers want trees that grow fast and produce a lot of wood. They want trees that resist disease and have high market demand.

5. **How do trees differ on the basis of wood, leaf shape, and leaf life?**

   Wood varies with the kind of tree. Some trees have soft wood; others have hard wood. Some trees have colorful wood; others don't. Some trees are easy to make into certain kinds of products, such as paper; others aren't.

   The two major shapes of leaves are needle and broadleaf.

   Leaf life may be year-round (evergreens) or seasonal (deciduous).

6. **Draw a tree and label the three major parts. Identify a product that we get from each part.**

   Note: The drawing should be similar to that on page 83 in the student text.

7. **What are the major activities in growing trees?**

   Growing trees involves protecting them from disease, insects, and pests and controlling fire. It also includes using the right cultural practices, such as thinning and prescribed burning in some places.

CHAPTER SELF-CHECK

1=B, 2=A, 3=G, 4=E, 5=H, 6=D, 7=F, and 8=C.

Chapter 8

GETTING PRODUCTS TO US

CHAPTER SUMMARY

People want food, clothing, and shelter in convenient, easy-to-use forms. People want to get their money's worth. Economical means that they get a good buy. Cost, durability, and source of products are important. People want their food to be wholesome.

Wholesome food is nutritious and free of dangerous substances. People want food that has been graded and is uniform in quality. Convenient foods are easy to prepare. They are placed in packaging appropriate for the product. People want environmentally safe products. They don't want products that pollute the environment or that produce a lot of waste. This requires good marketing procedures.

Marketing is the link that connects the producer with the consumer. A number of important steps are a part of the marketing process, known as marketing functions. These functions help assure safe food. Processing is used to preserve food. Canning, freezing, fermenting, drying, refrigeration, and other methods of preserving food are used.

People are consumers. They are the users of the goods and services that are produced. Consumers have preferences or demands that must be met.

INSTRUCTIONAL OBJECTIVES

The objectives of this chapter introduce students to marketing in the agricultural industry. The objectives are stated as questions in the student text.

Upon completion of chapter 8, the student will be able to:

1. Describe what people want in food, clothing, and shelter.
2. Explain how food, clothing, and shelter get to consumers in the forms that they want.
3. Explain food safety.
4. Describe the role of consumers.

INTEREST APPROACH, INSTRUCTIONAL STRATEGIES, AND TEACHING PLANS

Since all people are consumers, a wide range of experiences can be a part of the instruction with chapter 8. Students as consumers have certain preferences. This background serves as a base for the interest approach and the content of the chapter.

Have students read the introductory part of the chapter. Ask them to explain what is meant by the statement, "most crops must be changed to be the way we want them." Ask students if they would like to buy a live chicken or pig and prepare it into a meal.

Move from the interest approach into a presentation of the objectives. These are stated as questions on page 92 in the text.

Following the objectives, begin the content of the chapter. Have students read the chapter as homework or as supervised study in the class. Discuss the content of each section. Write key terms and definitions on the chalkboard. Invite students to share personal experiences they or their family members have had as consumers.

REVIEW AND EVALUATION

Review and evaluation often occur simultaneously. Review as well as evaluation shows areas where reteaching may be needed.

Use the "Reviewing" section at the end of the chapter. Have students read the section on "Main Ideas." Ask one or more students to orally summarize the "Main Ideas." Also have students complete the "Questions." This may be as homework or supervised study. Some teachers may prefer to have students orally answer them during the class discussion that concludes the chapter.

Evaluation can involve assessing the performance of students throughout the chapter, especially the "Questions" in the "Reviewing" section. Teachers may also want to use the "Chapter Self-Check" as an evaluation. A test may be constructed using the computer test item bank.

ADDITIONAL RESOURCES

Additional resources that may be useful include the book, *AgriMarketing Technology* (Interstate Publishers, Inc.). This book gives information on agricultural marketing from the perspective of the activities that people perform in marketing.

Information on marketing from the daily newspaper may also be useful, including the commodity report section.

ANSWERS

QUESTIONS

1. How do people want their food, clothing, and shelter?

People want their food, clothing, and shelter to be economical, wholesome, convenient and easy to use, and environmentally safe.

2. Why is the cost of things important?

People must be sure that they have enough money to pay for something. People want good value for their money.

3. What makes products durable?

Durable products resist wear and decay. Using good material and carefully manufacturing products helps to make them durable.

4. What do people look for in wholesome food?

Wholesome food must provide nutrition, be free of dangerous substances, and be graded and uniform.

5. Why do people want things that are convenient?

People are busy. They don't want to spend a lot of time preparing their food and clothing.

6. Why do people want environmentally safe products?

People want a good environment. They don't want to use products that create pollution when manufactured, such as toxic substances and wastes.

7. Briefly explain the marketing process.

Marketing is moving products from the producer to the consumer. It involves changing products into the forms that people want and keeping the products good to use. Marketing includes several marketing functions.

8. Why is food preservation important?

Food preservation keeps food from spoiling. Preservation keeps food safe to eat.

CHAPTER SELF-CHECK

1=C, 2=A, 3=B, 4=D, 5=E, 6=G, 7=H, and 8=F.

Chapter 9

USING SCIENCE IN AGRICULTURE

CHAPTER SUMMARY

Science is knowledge gained from experimentation. Experiments are tests or trials of something. The process that is used in experimenting to gain science information is known as the scientific method.

The scientific method involves the use of carefully designed experiments. Research is an important part of gaining new information. Research is the careful, systematic investigation for the purpose of discovering science. Research is also said to be answering questions. Researchers use the scientific method, which consists of five steps: identify the problem, gather data about the problem, develop possible solutions, test the solution(s), and evaluate the results.

Agriculture represents an exciting area for the use of science. Many new developments have come about through agricultural research. Exciting new research is underway that will impact the future in agriculture.

INSTRUCTIONAL OBJECTIVES

The processes of science in agriculture are explained in this chapter. The objectives contribute to the overall learning outcome. The text states the objectives as questions.

Upon completion of Chapter 9, the student will be able to:

1. Explain science.
2. List and define the major areas of science.
3. Describe how scientific investigations are conducted.
4. Explain the connections between agriculture and science.
5. List examples of the use of science in agriculture.

INTEREST APPROACH, INSTRUCTIONAL STRATEGIES, AND TEACHING PLANS

Instruction in chapter 9 lends itself to considerable creativity by the teacher and student. The natural inquisitiveness of students can be put to good use.

Begin the chapter by having students read the introductory section. Ask them to summarize the introduction and orally give examples of foods that they like. Have the students give examples of how the foods were developed or prepared.

Following the interest approach, present the objectives for the chapter. This may involve referring the students to the list of questions on page 106.

Begin the content of the chapter by having students read the various sections during class or as homework. Review the content of the sections through discussion methods. Use the chalkboard to list key terms and definitions.

Have students conduct the rag doll seed germination activity described in the chapter. This may be with corn, bean, or other seed. Different observations can be made of the germination trials by the students. Have students calculate their results and explain why the information is important and useful to a farmer.

Refer students to examples of research bulletins that give reports of experiments that have been carried out. Each student may be called on to discuss different research reports.

REVIEW AND EVALUATION

Review and evaluation can involve discussion and observation of the activities in this chapter. Review may involve using the "Reviewing" section, especially the "Main Ideas" at the end of the chapter. Review can also involve having students answer the "Questions" orally or in writing.

Evaluation may involve the use of the "Chapter Self-Check" and teacher-made tests. The performance of students on the "Exploring" activities will be useful in evaluation.

Reteaching will involve observations by the teacher during the review and evaluation. The "Agriscience Vocabulary" list may also be used in reteaching.

ADDITIONAL RESOURCES

Research bulletins, agricultural magazines with research reports, and similar materials can be used to enrich the chapter. In some cases, agricultural scientists who work in research may be available and can be used as resource persons in the class.

Seed and paper towels will be needed to carry out the rag doll seed germination test.

ANSWERS

QUESTIONS

1. **What are the major areas of science?**

   The major areas of science are physical science, life science, and social and behavioral science.

2. **How do areas of agriculture fit into the basic areas of science?**

   Agriculture involves producing living things; therefore, agriculture makes a lot of use of the life science. The practices in producing the plants and animals requires a lot of mechanics or physical science. Social and behavioral science gives information about the needs and interests of people.

3. **How are experiments conducted?**

   An experiment is used to test something. The testing must involve careful procedures in doing the experiment.

4. **Which variable in an experiment is measured?**

   The dependent variable is measured in an experiment.

5. **Why must experiments be carefully designed and conducted?**

   Experiments must be carefully designed and conducted if the results are going to be accurate and useful in making decisions.

6. **What is an agriscientist?**

   An agriscientist carries out research in different areas of agriculture.

7. **Where do agriscientists conduct their research?**

   Agriscientists do their research in laboratories and on farms or in any agricultural facility.

8. **What are some examples of applications of science in agriculture?**

   (Examples of science in agriculture are listed in the chapter. Students may list these or others that may be studied in class. A minimum of three should be listed.)

CHAPTER SELF-CHECK

1=H, 2=A, 3=E, 4=F, 5=G, 6=C, 7=B, and 8=D.

Chapter 10

SUSTAINING ECOSYSTEMS

CHAPTER SUMMARY

An ecosystem is the community of living and non-living elements. They work together in a system or cycle to ensure the survival of every living thing in the ecosystem. The living things form a biotic subsystem, while the non-living things form the abiotic subsystem.

Energy is the most important thing necessary to sustain an ecosystem. The energy is transferred in an ecosystem by a food chain. An ecosystem has producers, consumers, and decomposers. In some cases, the same food is needed by more than one organism. This forms a food web.

A community is made up of all of the living plants and animals in an area. The place in the community where one individual lives is its habitat. Habitats must provide the food, shelter, space, and other needs of the organisms.

Nature is always changing. Growth is always going on in an ecosystem. Survival of organisms requires that organisms adapt to their environment. Human families are a kind of small ecosystem. Families are continually changing when new babies are born and other members die.

Ecosystems are important in agriscience. Wild foods are not very important in meeting the needs of humans. The kinds of plants and animals that humans grow require attention so that their needs are met. Agriscience provides information to help in meeting these needs. Understanding the ecosystem requirements of domestic plants and animals helps agriscientists to be more productive in their work.

INSTRUCTIONAL OBJECTIVES

The overall purpose of chapter 10 is to help students understand important relationships of ecosystems to the production of food, clothing, and shelter. The objectives are listed in the student text as questions.

Upon completion of chapter 10, the student will be able to:

1. Define ecosystem.

2. Explain a food chain and how it is a part of the ecosystem.

3. Describe the relationships that exist in communities.

4. Explain ecological succession and the factors that are involved.

5. Explain biodiversity and its role in an ecosystem.

6. Describe how domestication and agriscience have affected the earth's ecosystems.

INTEREST APPROACH, INSTRUCTIONAL STRATEGIES, AND TEACHING PLANS

This chapter can be related to many features of the local community. It is likely that the best vehicles to teach the content will involve local community features and relationships.

The interest approach can begin by having the students read the introductory part of the chapter. Ask them to explain the communities that they are a part of and how these are systems. Also ask them to explain how the systems that we have must work properly. Open discussion will help prepare the students for the objectives.

Present the objectives by referring the students to page 120 in the text or listing them on the chalkboard. Following the "Objectives", begin the content of the chapter.

Students can benefit by reading the chapter for understanding. The reading may be homework or during supervised study in the classroom. Present each section of the chapter using discussion and the chalkboard to summarize key terms and concepts. Taking a short field trip to study an ecosystem will help students achieve the objectives. Have them name the different parts of the ecosystem and explain what they contribute to it. (A corner of the schoolgrounds or other place will likely have sufficient living and non-living elements to serve as an ecosystem for this study.)

REVIEW AND EVALUATION

Review and evaluation may involve several of the same techniques. Have students read the "Main Ideas" section at the end of the chapter and summarize its meaning. Also, the students may answer the "Questions" orally in class or in writing as homework.

Evaluation can include how well the students performed on the review as well as completion of the "Chapter Self-Check." Teacher made tests may also be used in evaluation.

Reteaching may involve a variety of techniques. What to reteach will be identified in evaluation. Some teachers use the "Agriscience Vocabulary" list at the beginning of the chapter for reteaching.

ADDITIONAL RESOURCES

Access to a small outside area convenient to the classroom will be beneficial in helping students understand the meaning of ecosystems and the elements that are involved.

This would likely be a good chapter to have students keep a journal. Understanding their community as an ecosystem is a good topic for writing in a journal.

ANSWERS

QUESTIONS

1. **What are the parts of an ecosystem?**

   An ecosystem is comprised of biotic and abiotic elements.

2. **What is a food chain? Describe how producers, consumers, and decomposers contribute to the survival of an ecosystem.**

   A food chain is the path that energy (food) travels in an ecosystem. Producers are the plants that use the energy from the sun to produce food. Consumers do not make food and get their energy from other sources—plants and animals. Decomposers are the organisms that live off of dead tissue. They help dead plants and animals decay.

3. **How does a habitat differ from a niche? Choose a living animal and use it as an example when answering this question.**

   A habitat is the place where a plant or animal naturally lives. A niche is the function a plant or animal fills in its habitat.

   (Any plant or animal can be selected by the student, with the explanations of habitat and niche carefully included.)

4. **What is ecological succession and how can it be affected?**

   Ecological succession is the natural change that occurs in an ecosystem. Humans can do things that speed up or slow down processes that are occurring in nature.

5. **What affect does biodiversity have on the ecosystems of the earth?**

   Biodiversity is when a wide variety of plants and animals exist together in an area. It gives each species more to eat and a better chance to survive.

CHAPTER SELF-CHECK

1=H, 2=E, 3=A, 4=C, 5=B, 6=G, 7=F, and 8=D.

Chapter 11

GROWING PLANTS

CHAPTER SUMMARY

Plants are important! Without plants, humans couldn't survive. Plants are necessary for all animal life. Animals get their nutrition from plants.

Plants also provide a lot of other benefits. They help keep the earth a good environment for people and all other living things. Many of the products people use in their daily lives come from plants.

The most important types of plants are food crops, ornamental plants, and forage and fiber crops. These plants are made into many products that are useful to people.

The way plants grow is important to the agriscientist. Plants begin growth in the embryonic stage and go through a juvenile stage before becoming mature plants. Mature plants have flowers and form fruit, seed, and other structures. Photosynthesis is the process plants use to convert energy from the sun into food. Plant reproduction involves the production of pollen, which is the male reproductive cell. The female reproductive organ in plants is the stigma. The transfer of pollen to the stigma is known as pollination or fertilization.

In order to grow properly, plants need light, temperature, nutrients, water, and gases from the air. People can help plants grow by providing needed nutrients and controlling pests that damage crops. Research has produced improved plants. Plant breeding programs have resulted in improved crops, such as high lysine corn, and new crops, such as triticale.

INSTRUCTIONAL OBJECTIVES

The objectives for this chapter are designed to help students understand the important role of plants. The objectives are listed on page 132 in the text as questions.

Upon completion of chapter 11, the student will be able to:

1. Describe how plants are important.
2. List the common types or groups of plants.
3. Explain how plants grow.
4. Name what plants need in order to grow.
5. Explain how people can help plants grow.

INTEREST APPROACH, INSTRUCTIONAL STRATEGIES, AND TEACHING PLANS

Students like living things. They like to know what makes them live and grow. This chapter can build on these motivations of students. The strategies and plans will vary with the students' needs and interests, as well as the agriculture found in the local area.

Begin the interest approach by having students read the introductory part of chapter 11. Ask students to explain what is meant by "...people have done more and more to manage plant growth." Of course, the content of the chapter is about providing good conditions for growing plants. Ask students to offer specific examples related to the local community.

Following the interest approach, present the objectives for the chapter. This can be by reviewing the questions on page 132 or listing the objectives on the chalkboard.

The content of the chapter should be covered following the objectives. Begin with the section on "The Importance of Plants." Have students read the section and provide an oral summary. Write key terms and definitions on the chalkboard. Follow a similar procedure with the remainder of the chapter. Bringing growing plants into the classroom can enrich the instruction. Explain how the growth of these plants is related to the content of the chapter.

REVIEW AND EVALUATION

The "Reviewing" section at the end of the chapter can be used in providing a review. First, have students read the "Main Ideas" section and orally summarize what they have read. Secondly, have students orally or in writing provide answers to the "Questions."

Evaluation can involve the review activities, as well as the "Chapter Self-Check." Teacher-made tests or a test constructed from the computer test bank can be used.

ADDITIONAL RESOURCES

Many additional resources can be used with this chapter. The amount of activity depends on the amount of time that is available. Students can make use of a greenhouse or school garden. Some students may have gardens at home as a part of their supervised practice.

Brochures and farm magazines that provide information about crops grown in the local area will be useful.

ANSWERS

QUESTIONS

1. **Are plants necessary for human survival? Why?**

   Plants are necessary for human survival because all food comes from plants in one way or another.

2. **What are the major groups of agricultural plants?**

   The major groups of agricultural plants are food crops, ornamentals, and forage and fiber crops.

3. **How do plants grow?**

   Plants grow by increasing the number and size of cells.

4. **Why do plants need water and light to grow?**

   Plants need water and light so that they can make food in a process known as photosynthesis.

5. **Does plant growth occur at the tips of stems or at ground level?**

   Plant growth occurs primarily at the tips of stems and roots.

6. **What happens when a plant reaches maturity?**

   Plants begin to produce flowers when they reach maturity. Later, the plant tissues begin to breakdown and die.

7. **What is pollination? How does it occur? What is the result?**

   Pollination is the fertilization process in plants. Pollen is transferred to the stigma. The result is the formation of seed or other structures, such as fruit, depending on the kind of plant.

8. **What are some improvements in plants made by scientists and growers?**

   Improvements to plants include those that grow in a variety of climates, produce bigger yields, and resist disease.

CHAPTER SELF-CHECK

1=H, 2=D, 3=G, 4=E, 5=B, 6=A, 7=F, and 8=C.

Chapter 12

RAISING ANIMALS

CHAPTER SUMMARY

Animals are important in many ways to people. They are used to provide food and clothing, as work animals, as pets, and as beautiful wildlife. Animal producers need to know the requirements of the animals they are producing. The quality of the products that we get relates to the care given the animals.

The products animals produce include meat, milk and dairy products, eggs, and wool. Each is produced by a specific type of animal. Meat is from beef cattle, swine, sheep, chickens, turkeys, and fish. Milk is primarily from dairy cattle, though goats and a few other species may be milked. Eggs are primarily from chickens.

Animals grow by increasing the size of muscle, bone, organs, and other body parts. Most growth is due to an increase in the number of cells. Hormones influence how growth takes place. Producers of animals need to provide the conditions to help animals grow efficiently. Nutrition is the kind and amount of feed as related to the needs of the animal. Six different nutrients are needed: carbohydrates, proteins, fats, minerals, vitamins, and water. Producers must provide the right environment and help prevent disease.

INSTRUCTIONAL OBJECTIVES

The objectives of chapter 12 are intended to help students understand the fundamentals of producing animals. The objectives are stated as questions on page 146 of the text.

Upon completion of chapter 12, the student will be able to:

1. Discuss how animals are important.
2. List and distinguish between the common types of animals.
3. Explain how animals grow.
4. Describe the requirements for animal growth.
5. Explain how producers can help animals grow.

INTEREST APPROACH, INSTRUCTIONAL STRATEGIES, AND TEACHING PLANS

Students usually have a high interest in animals. This interest should be used in providing the instruction. Begin the instruction with an interest approach.

One interest approach is to have students read the introductory part of the chapter. Ask them to name their favorite meat dishes. Ask them to describe how the meat is produced. Indicate that there are many important things to consider in raising animals.

Following the interest approach, present the objectives for the chapter. This may be by having a student read aloud the questions on page 146 or writing them on the chalkboard.

Cover the content of the chapter one section at a time. Write key terms and definitions on the chalkboard. First, review the meaning of domestication. This can be by calling on a student to explain domestication. Make a list of the products people get from animals on the chalkboard. List by each product the kind of animal that produces it. A field trip to an animal farm may be appropriate, especially if one of the students lives on a farm where animals are produced.

Explain how animals grow. This may involve reviewing the fact that animals are made of cells. Use a microscope to help students see a cell. Relate the importance of nutrition to animal growth by having students read and orally summarize this section of the chapter. After students have read the section on "Helping Animals Grow", use their input to summarize the major points on the chalkboard.

REVIEW AND EVALUATION

The "Reviewing" section of the chapter may be useful in both review and evaluation. Have students read the "Main Ideas" section and provide oral statements about what they have read. These statements are also useful in evaluating learning. Students can also provide the answers to the "Questions" at the end of the chapter orally or in writing.

Evaluation can involve using the "Chapter Self-Check", teacher-made tests, performance on the "Activities" at the end of the chapter, and other means.

ADDITIONAL RESOURCES

Brochures and magazines about the animals found in the local area may be useful. Students can prepare reports, construct posters or bulletin boards, and do other activities. Information about the kinds and numbers of animals in the local area will be useful.

ANSWERS

QUESTIONS

1. **What are the trends in meat consumption in the United States today?**

   Consumption of beef has declined. Consumption of pork, chicken, and turkey have increased.

2. **Why is lean meat a good food in the diet of humans?**

   Lean meat is an excellent source of nutrients for humans. Meat is especially high in protein.

3. **What nutrients are provided in relatively large amounts by meat products?**

   Meat products provide protein, which contains amino acids. The amino acids are essential for growth and rebuilding the body when it is injured.

4. **How do animals grow?**

   Animals grow by increasing the size of muscle, bone, organs, and other body parts.

5. **What factors affect animal growth?**

   Animal growth is affected by the nutrients they receive, as well as their protection and space available for them.

6. **How do environmental factors affect animal growth?**

   Environment greatly influences growth. Animals need protection from cold and hot weather.

7. **What nutrients do animals need to grow and be healthy?**

   Animals need the following nutrients to grow and be healthy: carbohydrates, proteins, fats, minerals, vitamins, and water.

CHAPTER SELF-CHECK

1=B, 2=H, 3=D, 4=F, 5=E, 6=C, 7=A, and 8=G.

Chapter 13

KEEPING ANIMALS AS PETS

CHAPTER SUMMARY

Humans have kept animals as pets for thousands of years. A pet is an animal that is domesticated, treated with kindness, and kept as a companion. It is the same as a companion animal. Pets are beneficial to people in many ways. Pets should be carefully chosen and only after a person has made a commitment to be a good care giver. Choosing a pet also involves knowing about the laws that apply. Exotic animal and leash laws are common.

Pets must have a proper environment. A container where a small pet may be kept is known as a vivarium. Those for aquatic pets are known as aquariums, while those for terrestrial pets are known as terrariums.

The common types of pets are dogs, cats, birds, fish, small mammals, and reptiles and amphibians. In some cases, people may have horses, llamas, or other large animals as pets.

Pets help people. Pet facilitated therapy (PFT) is used to help sick people heal. Service dogs help in many ways, such as herding cattle or assisting people who have disablities. Seeing-eye dogs (also known as leader dogs for people who are severely visually impaired) and signal dogs (for people who have hearing impairments) are very important in helping people assume productive roles in society.

INSTRUCTIONAL OBJECTIVES

Since most young people like pets, the students should be motivated for the content of this chapter. The objectives are stated as questions on page 160 of the student text.

Upon completion of chapter 13, the student will be able to:

1. Define pet and describe how pets are important to people.
2. List what to consider before getting a pet.
3. Describe the care requirements of pets.
4. List the important types of pets.
5. Explain how companion animals are helpful.

INTEREST APPROACH, INSTRUCTIONAL STRATEGIES, AND TEACHING PLANS

The overall plan for this chapter can focus on the pets of the students in the class. Small animals can also be brought to school as a part of this chapter. (Caution: All animals should be properly handled. Check on the regulations of the school before bringing animals to school.)

Begin the interest approach by having students read the introductory section of the chapter. Develop a list of the pets they own on the chalkboard. Have several students give background information about their pet, such as the pet's name and how long they have had them.

Following the interest approach, present the objectives for the chapter. This may be by having the students review the questions on page 160 or writing the objectives on the chalkboard.

Cover the content of the chapter by having students read the sections of the chapter and orally summarize what they have read. Write key terms and definitions on the chalkboard. Be sure to use the full definition of "pet," as presented on page 161. Invite the local animal control officer or an animal protection shelter representative to serve as a resource person and describe leash laws and how animals are handled. Carefully review the "Questions to Answer" when deciding about a pet.

Construct a vivarium for a pet in the classroom. Many teachers prefer to use fish and have an aquarium. Some schools teach aquaculture classes and are able to use the fish in many ways in the instructional program.

REVIEW AND EVALUATION

Use the "Reviewing" section at the end of the chapter for both review and evaluation. Have students read the "Main Ideas" and orally or in writing answer the "Questions."

Use the "Chapter Self-Check", teacher-made tests, and participation in class activities in evaluation. Reteach on the basis of learner achievement. The "Agriscience Vocabulary" list on page 160 may be used in review, evaluation, or reteaching.

ADDITIONAL RESOURCES

Many resources can be used to teach about pets and companion animals. Books, brochures, and magazines on pets and pet care may be useful to the teacher and students.

Facilities to set up a vivarium in the class room will be needed if to do so is a part of the instructional strategy.

ANSWERS

QUESTIONS

1. **What are the benefits companion animals provide for their human care givers?**

   Companion animals help people cope with stress and stay healthy. Pets are often soothing and relaxing to people.

2. **What must a human care giver provide so that their companion animal will get its basic needs met and live a contented, comfortable life?**

   Care givers need to provide a good environment for the pet. They need to give it food, water, and shelter. People also need to look after the health of the animal, which may include vaccinations to prevent disease.

3. **What are the major categories of companion animals? Give examples in each category that you list.**

   Dogs, cats, birds, fish, small mammals, and reptiles and amphibians may be kept as companion animals. (The examples given may be based on what the students or their families have for pets. Any examples from the text will be appropriate.)

4. **Imagine you are about to choose a pet. What factors must you consider before you bring a pet home?**

   (This answer will likely be built around the list of questions on pages 164 and 165. These questions cover the important factors in choosing a pet.)

CHAPTER SELF-CHECK

1=G, 2=F, 3=B, 4=C, 5=E, 6=D, 7=H, and 8=A.

Chapter 14

USING POWER MACHINERY

CHAPTER SUMMARY

Machines are devices that do work. Several machines may be combined into one mechanical device, known as machinery. Power is an important part of machinery.

Power is the rate at which work is done. The major sources of power are electric motors and internal-combustion engines. Draft animals aren't used much for power. Steam is used in some situations. Some power sources are mobile; others are stationary. Tractors are mobile sources of power.

Automation is using machines to do work that requires making decisions about the work. Controls are used in automation to direct the machinery to perform certain work. Computers are used on machines to process information at a high rate of speed. Robots are computer-operated machines that can do monotonous routine work.

Agricultural machinery is used in all areas of the agricultural industry. Supplies and services, products processing, and production of crops and animals involve the use of power machinery. Machinery is used to till land, plant crops, control pests, and harvest crops. Automated feeding equipment, milking and egg-gathering equipment, and other machinery is used with animal production.

INSTRUCTIONAL OBJECTIVES

The objectives for this chapter are stated as questions on page 174 in the text. The instructional objectives are stated here.

Upon completion of chapter 14, the student will be able to:

1. Explain automation and power machinery.
2. List the kind of machinery that is commonly used in agriculture.
3. Describe why science is important in power machinery.

INTEREST APPROACH, INSTRUCTIONAL STRATEGIES, AND TEACHING PLANS

Some students will be very interested in agricultural machinery. They will be able to relate the color of paint that is used on different brands of machinery, as well as give models and horsepower ratings. This interest can be promoted in the chapter.

The interest approach can involve having the students read the introductory part of the chapter. Ask them to name ways power machinery has changed how food, clothing, and shelter are produced. Develop a list on the chalkboard.

Move from the interest approach into the objectives. The students can be referred to page 174 in the text or the objectives can be presented using the chalkboard or overhead projector.

The content of the chapter can be covered sequentially through the chapter. Begin with "Machines, Power, and Automation." These are fundamental concepts of physics that underlie the design and use of machinery. A field trip to a farm machinery dealership, farm show, or farm with machinery can help students see machinery. In some cases, the machinery can be observed in operation. (Caution: Be sure that safety practices are reviewed before taking the field trip and being around machinery that is operating. Be sure to follow school procedures on such activities.)

This chapter presents a good opportunity for students to prepare posters or bulletin boards that show pictures of machinery. The posters should focus on machinery for a particular crop, such as corn or wheat, or machinery that performs specific functions, such as planting or harvesting.

REVIEW AND EVALUATION

Use the "Reviewing" section at the end of the chapter. Have students read the "Main Ideas" section and orally report on the summary. Also have the students answer the questions at the end of the chapter. This may be in writing during supervised study or as homework or orally during class time.

Evaluation can involve the review activities, as well as the "Chapter Self-Check", teacher-made tests, and other means.

Reteaching can be based on student performance in the review, as well as evaluation.

ADDITIONAL RESOURCES

The instruction can be enriched with brochures from farm machinery dealerships, farm magazines with advertisements about farm machinery, and in other ways.

Additional information on the application of science in power machinery is found in *Introduction to World AgriScience and Technology* (Interstate Publishers, Inc.), *Physical Science Applications in Agriculture* (Interstate Publishers, Inc.), and *Mechanical Technology in Agriculture* (Interstate Publishers, Inc.).

ANSWERS

QUESTIONS

1. **Why is machinery important in agriscience?**

    Machinery is important in agriscience because it can do monotonous, hard labor work faster than humans.

2. **What are the two main sources of power with agricultural machinery? Compare the two.**

    The two main sources of power are electric motors and internal-combustion engines. Electric motors convert electrical energy into mechanical power. Internal-combustion engines convert heat into power.

3. **Why is automation beneficial to people?**

    Automation is used to do jobs that are dangerous and complex. It saves people from having to do the work.

4. **Distinguish between mobile and stationary power. Why is mobile power important?**

    Stationary power isn't movable, while mobile power is movable and can go where it is needed. Mobile power allows tractors to operate in fields pulling implements.

5. **How is tillage used in crop production?**

    Tillage is plowing the soil. It is used to make a seed bed and destroy weeds.

6. **Distinguish between planters and drills.**

    Planters plant seed in rows; drills do not plant seed in rows.

7. **Why can't the same harvesting equipment be used for all crops?**

    Crops are very different in how they grow and the forms of their products.

8. **Some people like farm machinery because it involves science. Explain how science is involved.**

    Science is used to design machinery. Physics is one of the most important areas of science used in developing machinery.

CHAPTER SELF-CHECK

1=B, 2=E, 3=A, 4=D, 5=G, 6=C, 7=F, and 8=H.

Chapter 15

HELPING PLANTS AND ANIMALS GROW

CHAPTER SUMMARY

People expect a lot from plants and animals. Improving how plants and animals grow requires knowledge of their environmental and nutrient needs. They must also be protected from pests and given proper care. People usually get a lot of benefit from helping plants and animals grow.

The benefits of improving plant and animal growth include bigger yields, better quality, easier work in harvesting and other areas, and more profit to the owner. With animals, well being is a definite consideration. Animal well being is caring for animals so that they don't suffer.

Biotechnology is used to create or improve plants and animals. It may involve changing the plant or animal or creating a better environment for it. Cloning is used to reproduce plants and animals by using a small piece of tissue or a few cells. Tissue culture is increasingly being used with plants.

People are also concerned about helping plants and animals adjust to changes in the environment. Growth chambers are being used in agriscience to study the effects of certain changes. Hormones are being used to cause increased production or prevent unwanted growth.

Genetic engineering is a popular and controversial topic. It involves changing the genetic makeup of a plant or animal. Some people are concerned that this may damage the environment or pose dangers to humans. Very few genetically-engineered crops are being used.

INSTRUCTIONAL OBJECTIVES

The objectives for this chapter are intended to introduce students to biotechnology and related areas. The objectives are stated as questions on page 188 in the text.

Upon completion of chapter 15, the student will be able to:

1. List ways people help plants and animals grow.
2. Explain the benefits of helping plants and animals grow.
3. Describe ways biotechnology is used in agriscience.
4. Discuss uncertainty in the use of biotechnology.

INTEREST APPROACH, INSTRUCTIONAL STRATEGIES, AND TEACHING PLANS

Students may have heard a variety of opinions about the ways people go about helping plants and animals grow. In particular, they may know that biotechnology is sometimes controversial. This chapter represents a good opportunity to allay some of the fears of students.

The interest approach can involve having students read the introductory section of the chapter. Ask the students to explain how helping plants and animals grow might result in damage to the environment. Indicate that this chapter is intended to help them understand the importance of using biotechnology in a responsible way.

Following the interest approach, review the objectives for the chapter. This can be done by asking students to read the questions on page 188.

Cover the content of the chapter beginning with "How Plants and Animals Need Help" and moving sequentially through the chapter. Have students read the sections. Write key terms and definitions on the chalkboard. Refer the students to research reports or magazine articles on biotechnology.

Small group discussions or debate procedures may be used to conclude the chapter. Students need to realize the consequences of biotechnology, particularly genetic engineering. They also need to understand the importance of laboratory testing before a new life form is released. Students also need to understand that biotechnology is essential for continued progress in feeding the world's growing population.

REVIEW AND EVALUATION

Use the Reviewing section at the end of the chapter. Begin by having students read and orally summarize the "Main Ideas." Afterward, have them orally or in writing answer the questions. The performance of students on the review may serve in evaluation.

Evaluation of the achievement of the objectives may include using the "Chapter Self-Check", teacher-made tests, and other means. Reteaching should be based on student achievement of the objectives.

ADDITIONAL RESOURCES

Teachers may wish to use materials on biotechnology, especially as related to agriculture. References include *Biological Science Applications in Agriculture* (Interstate Publishers, Inc.) and *Introduction to World AgriScience and Technology* (Interstate Publishers, Inc.). The *Biotechnology* and *Discover* magazines often have articles that relate to agriculture and the use of genetic engineering and other processes. Research reports from agricultural experiment stations may also be useful.

ANSWERS

QUESTIONS

1. **How do people help plants and animals grow?**

   People help meet the needs of plants and animals. This includes a good environment, proper nutrients, protection from pests, and good care.

2. **Why should people help plants and animals grow?**

   People get benefits from helping plants and animals grow. These include bigger yields, better quality, easier work, and more profit.

3. **Why is animal well being important?**

   Animal well being is important because it involves caring for animals so that they don't suffer. People don't want animals to be abused. They also want animals to be healthy.

4. **How does biotechnology help plants and animals live better?**

   Biotechnology helps plants and animals by improving their conditions of life.

5. **How does genetic engineering change living things?**

   Genetic engineering changes the natural characteristics of a plant or animal. Genes may be transferred from one plant or animal to another.

6. **How do we make good decisions about the use of biotechnology?**

   People need to know the facts about biotechnology. This means that they must keep informed about research and new developments.

CHAPTER SELF-CHECK

1=B, 2=F, 3=D, 4=H, 5=A, 6=G, 7=E, and 8=C.

Chapter 16

KEEPING A GOOD ENVIRONMENT

CHAPTER SUMMARY

People want to live in a good place. They want the earth to provide a good environment. Studying the environment is known as environmental science. Ecology is the science of how living things interact with each other and with their environment.

Most living things depend on other living things for survival. This interdependent relationship serves different purposes with different organisms. An ecological system (ecosystem) is the community in which plants and animals live and share the same area. Some of the plants and animals compete with each other for food, space, or in other ways. Nature is continually changed and never really in balance. Succession is the change that is underway in an ecosystem which involves the natural return of plants and animals to an area.

In order to keep a good environment, conservation is needed. Five important areas of conservation are soil conservation, water conservation, forest conservation, wildlife conservation, and mineral conservation.

Protecting the environment is everybody's business. Three main ways are used. First, when we use a resource, we should use it as fully as possible, thus reducing the amount we need. Second, when we have finished using a resource, we should recycle it if possible. Third, conservation measures help to provide resources for future generations. Government agencies, private organizations, and individuals are active in conservation.

INSTRUCTIONAL OBJECTIVES

The instructional objectives for the chapter are stated as questions on page 200 in the text. Students may wish to refer to the questions.

Upon completion of chapter 16, the student will be able to:

1. Define environmental science.
2. Explain how environmental science and ecology are related.
3. Describe important principles of ecology.
4. Explain how conservation and environmental science are related.
5. List the types of conservation used in environmental science.
6. Identify who is responsible for environmental science and conservation.

INTEREST APPROACH, INSTRUCTIONAL STRATEGIES, AND TEACHING PLANS

A variety of procedures can be used in teaching chapter 16. One is to begin with an interest approach. A suggested approach that involves students with the content is to have them read the introductory part of the chapter. (Reading the material helps develop reading skills, as well as increases internalization of the content.) Ask them to offer suggestions on what can be done so that there will be resources to meet the needs of future generations. The resources involved may need to be identified.

Following the interest approach, present the objectives to the students. Have them read the questions on page 200 or write the objectives on the chalkboard.

Begin the content of the chapter by having them read the section that defines environmental science. Call on one or more students to offer their explanations of environmental science. Write key items on the chalkboard. Move to other sections in the chapter. Define ecology and discuss "interdependent relationships." Move into ecology and the environment. Discuss competition, balance of nature, and succession. After the students have read the section of the chapter on conservation, list the five areas on the chalkboard. Ask students to describe the areas. List key concepts associated with each on the chalkboard. Have students read the three main ways of protecting the environment. Call on different members of the class to explain each way of protecting the environment that is presented. Conclude the chapter with the section on "Responsibilities for the Environment." Invite a representative from a government agency and/or private organization to serve as a resource person in class to describe the role of conservation.

REVIEW AND EVALUATION

Use the "Reviewing" section at the end of the chapter. Students should read and orally summarize the "Main Ideas." Following, discussion have students answer the questions. This can be in writing as homework or supervised study or orally during class. The review activities can also be used as evaluation.

Use the "Chapter Self-Check" in evaluation. Teacher-made tests may also be used. Reteaching can be based on the findings of the evaluation.

ADDITIONAL RESOURCES

Information about conservation in the local area may be helpful. The Soil Conservation Service or other government agency may have useful materials that relate to your area. Also, brochures, magazines, and other publications may be helpful. Some teachers use the magazine entitled *EARTH: The Science of Our Planet.*

ANSWERS

QUESTIONS

1. **What is environmental science? Why is it important?**

   Environmental science is the study of the environment. It is important because it provides information to help with conservation.

2. **What is an interdependent relationship? Give an example.**

   An interdependent relationship is when living things depend on other living things in order to survive. An example is when animals eat plants for food.

3. **What is succession? How does it work?**

   Succession is the natural return of plants and animals to an area cleared by fire or humans. Grasses and small shrubs grow first. Small rodents and other animals live in the grass. After a time, larger shrubs and small trees grow. These are followed by large trees and large animals.

4. **What are the types of conservation? How are they related?**

   The types of conservation are: soil conservation, water conservation, forest conservation, wildlife conservation, and mineral conservation. They are related because damage to one also results in damage to another. For example, soil erosion may also pollute water with soil particles.

5. **What are three main ways we can protect the environment?**

   The three main ways of protecting the environment are:

   (1) When we use a resource, we should use it as fully as possible, thus reducing the amount we need.

   (2) When we have finished using a resource, we should recycle it if possible, so it can be

reused and reduce the need to harvest new resources.

(3) When we follow conservation measures, our soil, water, forests, wildlife, and mineral resources will be left for future generations.

6. **Briefly describe the three groups of people responsible for protecting the environment.**

   The groups responsible for protecting the environment include government agencies, private organizations, and individuals. Several federal agencies are included, such as the Soil Conservation Service. Private organizations are associations of people who have certain interests, such as the National Audubon Society. Individuals can act on their own to help conserve resources.

CHAPTER SELF-CHECK

1=E, 2=D, 3=A, 4=F, 5=B, 6=G, 7=H, and 8=C.

Chapter 17

KEEPING OUR SOIL

CHAPTER SUMMARY

The soil is the outer covering of the earth's crust that contains nutrients plants need to grow. It is formed from parent material by a process known as weathering. Soil formed from material that is moved about by wind is known as loess soil, while that moved about by water is alluvial soil, and that by glaciers is known as glacial soil.

Digging straight down into the soil for five or six feet reveals a soil profile, which shows the different layers of the soil. A layer is known as a horizon. The topsoil is the A horizon, while the subsoil is the B horizon, and the third layer is the C horizon, which is parent material.

Soil is composed of organic matter (which is decayed plants and animals), minerals, air, and water. Mineral materials are of three kinds: sand, silt, and clay. The percentages of these materials determine the soil texture. Soil also contains nutrients. The primary nutrients are nitrogen, phosphorus, and potassium. The secondary nutrients are calcium, sulfur, and magnesium. Micronutrients or trace elements include iron, manganese, zinc, copper, boron, molybdenum, and chlorine. The primary nutrients are needed in larger amounts than the secondary nutrients. Only very small amounts of the micronutrients are needed for plant growth. The potential hydrogen in soil is known as pH. pH is measured on a scale of 0 (acid) to 14 (alkaline), with 7.0 being neutral. Soil testing is used to determine the nutrients in the soil and recommend fertilizers for certain crop uses.

Soil is lost by erosion. Wind erosion and water erosion cause the greatest losses. Water erosion includes three types: sheet erosion, rill erosion, and gully erosion. Soil is conserved by using it for its best use, known as land capability classes. Conservation practices include using contour plowing, strip-cropping, terracing, cover crops, no-till or minimum-till farming, crop rotation, grassed waterways, and silt fences.

INSTRUCTIONAL OBJECTIVES

The objectives for this chapter are designed to help students understand the formation, components, and conservation of soil. The objectives are stated as questions on page 214 of the text.

Upon completion of chapter 17, the student will be able to:

1. Describe how soil is formed.
2. List the components of soil.
3. Describe how land is classified.
4. Explain how soil is lost.
5. Describe conservation practices that can be used with soil.

INTEREST APPROACH, INSTRUCTIONAL STRATEGIES, AND TEACHING PLANS

Many students will have some familiarity with soil and soil conservation. Use this as a base to build on in the instruction. The content of the chapter can be more relevant to students if the soil conditions in the local area are a part of class discussion.

Begin the interest approach by having students read the introductory part of the chapter. Ask them to explain why everything we have depends on the soil in one way or another. Also, ask the students to discuss why people are responsible for seeing that the soil is not depleted.

Move from the interest approach into a presentation of the objectives. The objectives are stated as questions on page 214 in the student text.

The content of the chapter is sequenced so that instruction and learning build on prerequisite learning. Begin with the section on "How Soil is Formed" and move to the other sections. Have students read the sections. Use the chalkboard to outline the content of the chapter with input from the students.

Take a short field trip to observe a soil profile. A ditch cut through soil or bank left by construction may be a good place. Try to identify the horizons.

REVIEW AND EVALUATION

Use the "Reviewing" section to help with the review process. Begin with the "Main Ideas." Have students read the "Main Ideas" section and orally summarize the content. The students can also complete the "Questions" section.

Evaluation can involve the review process, as well as the "Chapter Self-Check", teacher-made tests, and other means. Some teachers may prepare soil or land judging teams for competition. This chapter might be a good time to introduce land judging and use it to review, evaluate, and reteach the content of the chapter.

ADDITIONAL RESOURCES

Information about soils in the local area will be helpful. The Soil Conservation Service will have useful materials. Brochures and reports of soil surveys will be available from the state agricultural experiment station or Cooperative Extension Service. Materials on land judging will likely be available from the Cooperative Extension Service.

ANSWERS

QUESTIONS

1. **How is soil formed?**

   Soil is formed from parent material. Weathering breaks down rock and other materials.

2. **Explain the difference between primary nutrients, secondary nutrients, and trace elements.**

   Primary nutrients are the nutrients that are most needed for plant growth. Secondary nutrients are needed in smaller amounts than primary nutrients. Trace elements are needed by plants in very small amounts.

3. **What are the parts (components) of the soil?**

Soil is made of mineral matter, organic matter, air, and water.

4. Briefly describe some different types of soil erosion.

The two major types are wind and water erosion. There are three types of water erosion: sheet, rill, and gully. Sheet erosion is usually on nearly level land that may be covered with water. When the water runs off a thin layer of soil is carried with it. Rill erosion is when tiny channels get washed in plowed land. These channels may develop into gullies if not controlled. Gully erosion is when large, deep, washed out areas appear on hillsides.

5. Why is the use of land capability classes important?

Most land is best suited for certain purposes, known as land capability classes. These classes state the best use for the land. Land should be used appropriately.

6. Briefly describe the conservation practices discussed in this chapter.

The conservation practices in the chapter are contour plowing, strip-cropping, terracing, cover crops, no-till or minimum-till, crop rotation, grassed waterways, and silt fences. Contour plowing is running the rows of crops across the slope of land on a contour with the slope. Strip-cropping is alternating strips of row crops and cover crops on hillsides. Terracing is making deep furrows around the contours of the land. Cover crops are crops planted to protect the soil from erosion. No-till or minimum-till farming is planting seed with little or no plowing the land. Crop rotation is changing from a row crop one year to a cover or legume crop the next year. Grassed waterways are areas where water runs that have been planted to grass. Silt fences are plastic strips held in position by stakes or posts much like a fence to catch any soil that may be washed so that it doesn't go into streams.

CHAPTER SELF-CHECK

1=G, 2=A, 3=F, 4=D, 5=C, 6=B, 7=E, and 8=H.

Chapter 18

KEEPING OUR WATER

CHAPTER SUMMARY

Water is important in many ways. Humans use it for many purposes. The earth has a lot of water; however, much of it isn't in a form that can be used. Water can also be damaged by pollution. Water is not used up; it is used over and over.

Water is a simple substance made of hydrogen and oxygen (H_2O). Two atoms of hydrogen and one atom of oxygen combine to form one molecule of water. Water may contain many different substances. Distillation is used to produce water that doesn't have minerals, gases, and pollutants.

Three forms of water are found naturally on the earth: solid, liquid, and gas. The solid state is ice, while liquid is water and gas is vapor, known as humidity.

Water quality varies widely. Water quality is the condition of water for a particular use. The most common way of classifying water is freshwater, brackish water, and saltwater. The amount of salt in water is measured as parts per thousand (PPT). Freshwater has less than 3 ppt salt. Saltwater has 16.5 ppt or more of salt. Brackish water is a mixture of freshwater and saltwater, such as where a stream runs into an ocean.

The water cycle is nature's way of purifying water. Water from the surface soaks into the ground and forms groundwater. It is pumped out of the ground and used. Water on the surface of the land evaporates into the atmosphere and forms clouds. The clouds provide precipitation that falls to the earth and repeats the cycle. Wetlands are a part of the water cycle. Wetlands are lowing lying areas where water often stands. It soaks into the earth and undergoes cleaning for using again.

Water is used in many ways: human consumption, recreation, agriculture, wildlife habitat, and industry. Water can be protected by avoiding waste, avoiding pollution, cleaning up polluted areas, and using water control structures.

INSTRUCTIONAL OBJECTIVES

The objectives for the chapter are stated as questions on page 228 in the text. The instructional objectives for the chapter are listed here.

Upon completion of chapter 18, the student will be able to:

1. Explain the nature of water.
2. Describe water quality.
3. Explain the water cycle.
4. Describe how water is lost.
5. Define wetland and tell why wetlands are important.
6. List ways of protecting water resources.

INTEREST APPROACH, INSTRUCTIONAL STRATEGIES, AND TEACHING PLANS

This chapter expands the concept that most people have of water. It presents water conservation in a way that may be new to students. The stress in teaching needs to be on using water and not wasting or polluting it.

Begin the chapter with an interest approach. One strategy is to have the student read the introductory section of the chapter. Ask them to explain how water is used over and over. It's somewhat like two people taking a bath in the same water, but on a much larger scale!

Move from the interest approach into the objectives. These are stated as questions on page 228.

Cover the content of the chapter sequentially and on a section by section basis. Have students read each section. Use class discussion with key terms and concepts written on the chalkboard.

If convenient, set up a water distillation apparatus in the classroom. Use a water test kit and determine water chemistry before distillation and test the water afterward. Use a sample from a nearby stream or pond for distillation. Ask students if they would like to drink the water directly from the pond or after distillation. Remind students that distillation is one way of purifying water.

REVIEW AND EVALUATION

Use the "Reviewing" section at the end of the chapter. Have students read the "Main Ideas" and orally report what they read. Have them answer the "Questions" orally in class or in writing as homework or during supervised study.

Evaluation can involve the review activities, as well as the "Chapter Self-Check" and teacher-made tests.

ADDITIONAL RESOURCES

Information about water in the local area will be useful. A field trip to a municipal waste water treatment facility would be useful.

Using water test kits and microscopes to test samples of water should help students understand water chemistry, water quality, and water biology.

ANSWERS

QUESTIONS

1. Why is water chemistry important?

Water chemistry is the composition of water. It determines how water can be used. Some water can kill plants and animals if it is contaminated.

2. Which physical state (form) of water is most useful? Why?

Liquid water is most useful. It is the form people drink, use to grow plants and animals, use in factories, and in other ways.

3. Why is water quality important?

Water quality is the condition of water for a particular use. The quality of water may limit how it can be used.

4. Draw the water cycle. Include features of your local community in your drawing.

(The drawing should be similar to that on page 234 in the text.)

5. What are the important uses of water? Why are these important?

The most important uses of water are human consumption, recreation, agriculture, wildlife habitat, and industry.

6. What can people do to help protect water?

People can protect the water by avoiding waste and pollution, cleaning up polluted areas, and use water control structures.

CHAPTER SELF-CHECK

1=F, 2=G, 3=H, 4=B, 5=C, 6=E, 7=D, and 8=A.

Chapter 19

KEEPING OUR AIR

CHAPTER SUMMARY

Air is everywhere! It surrounds the earth and is made of gases, moisture, and solid particles. Nitrogen and oxygen are the major gases in the air. Every time air is used it is changed. The air also forms a protective layer above the earth known as the ozone layer (O_3). This layer filters out harmful radiation from the sun. It is being damaged by chlorofluorocarbons (CFCs). Air contains moisture or water vapor that is known as humidity. Solid particles in the air are known as particulate. These include dust, ash, soot, and other tiny solids.

Air is important because it supports life. Plants and animals need oxygen from the air in order to live. It protects the earth and provides moisture as precipitation for plants and animals. Air provides recreation and is used for communication and transportation.

Air is damaged by pollution. This results when wastes get into the air. Particulate is measured as the number of particles per cubic inch. Air over oceans may have 15,000 particles per cubic inch, while that over land may have 100,000 particles per cubic inch. Polluted air near big cities has far more particles in it. Smog, oxides, and radioactive and toxic wastes also pollute the air.

Humans must protect the air by avoiding pollution.

INSTRUCTIONAL OBJECTIVES

The objectives for the chapter are stated as questions on page 242. The instructional objectives for the chapter are listed here.

Upon completion of chapter 19, the student will be able to:

1. Describe air.
2. List why air is important.
3. Explain the atmosphere.
4. Describe how the air is damaged.
5. Name ways people can help protect the air.

INTEREST APPROACH, INSTRUCTIONAL STRATEGIES, AND TEACHING PLANS

The procedures used in teaching air should relate to situations in the local community. Air quality and sources of pollution vary. These can be used to make the instruction have local relevance.

The interest approach can involve having the students read the introductory part of the chapter and study the photograph at the bottom of page 241. Ask them how they would like to live near the factory. How would they like to breathe the air?

Present the objectives for the chapter using the list of questions on page 242 or by writing the objectives on the chalkboard.

Cover the content of the chapter in sequence with sections in the text. Begin with the section on "Air Science", having students read each section followed by discussion and writing key terms and concepts on the chalkboard.

One approach to illustrate particulate in the air is to remove a filter from a heating and air conditioning system. Observe the dust that has collected. View the particulate with magnification. Remind students that these particles are in the air that they breathe and that filters should be kept clean so that they work properly in cleaning the air in our homes, schools, and offices.

REVIEW AND EVALUATION

Use the "Reviewing" section at the end of the chapter. Have students read "Main Ideas" and orally summarize the content. Also have students answer the questions orally in class or in writing as homework or during supervised study.

Evaluation can involve the use of the "Chapter Self-Check", teacher-made tests, and other means. Reteaching would be based on the findings during review and evaluation.

ADDITIONAL RESOURCES

Use the daily newspaper to get information from the weather report on the air quality. Students can keep records of the air quality index for several days and study the trends that develop.

Other materials on emissions and pollution may be useful. The Environmental Protection Agency or a local office on environmental protection may be able to provide useful materials.

ANSWERS

QUESTIONS

1. **What is the atmosphere? What levels are included?**

   Atmosphere is the air above the earth. It is in layers, known as the troposphere, stratosphere, mesosphere, and thermosphere.

2. **What is found in the air? Which of these are pollution?**

   The air contains gases, moisture, and solid particles. Any thing that gets into the air and changes it is pollution. Most people think of pollution as being solid particles or particulate. Others do not include moisture as air.

3. **Why is the ozone layer important? How is it damaged?**

The ozone layer is the upper level of the stratosphere that is formed of a kind of oxygen known as ozone. It filters out harmful radiation from the sun. Certain kinds of gases damage the ozone layer, such as the chlorofluorocarbons (CFCs).

4. What are the important things air does for the earth?

Air supports life, protects the earth, and provides moisture back to earth as precipitation. The air also makes communication and air transportation possible. Air is sometimes used for recreation.

5. How can damage to the air be prevented?

People can follow steps that reduce air pollution. Government laws have established air quality standards. Individuals can do little things that make a big difference.

CHAPTER SELF-CHECK

1=D, 2=C, 3=B, 4=E, 5=A, 6=F, 7=H, and 8=G.

Chapter 20

KEEPING OUR WILDLIFE

CHAPTER SUMMARY

Wildlife is all non-domesticated plants and animals. They are found in nearly all places and vary in many ways. Most people think only of game animals (those that are hunted for sport), but wildlife includes trees, insects, birds, fish, and other plants and animals.

Some wildlife are extinct and others are endangered or rare. Extinct wildlife no longer live on the earth. Endangered wildlife is wildlife that is threatened with extinction. Rare wildlife exist in small numbers, usually in protected areas.

People can enjoy wildlife in many ways. Some people use it for sport hunting and fishing. Others use it for food. A lot of people watch wildlife. People like to see the wonders of nature.

Many kinds of wildlife are found. They can be classified into scientific taxonomies, such as plants and animals. Wildlife can be classified by habitat, such as aquatic or terrestrial.

Wildlife management is how people can use the natural environment to help wildlife live and grow. This requires understanding various habitat requirements. Animal wildlife need food, water, cover, and range and territory. Plant wildlife must have an environment that is appropriate to their needs. People can do a lot to help wildlife live and grow.

INSTRUCTIONAL OBJECTIVES

The objectives for the chapter are listed as questions on page 254 of the text. The instructional objectives are listed here.

Upon completion of chapter 20, the student will be able to:

1. Define wildlife and give examples.
2. Explain why wildlife are important.

3. Describe important wildlife management principles.
4. Define endangered wildlife.
5. List ways of protecting wildlife.

INTEREST APPROACH, INSTRUCTIONAL STRATEGIES, AND TEACHING PLANS

Wildlife is an area of high interest to many students, particularly those who hunt and fish. Use this interest as a base for the instruction.

The interest approach can involve having students read the introductory part of the chapter. Ask them to orally summarize what they have read. Ask about the meaning of, "Wildlife may not be so happy to see people."

Present the objectives for the chapter using the list of questions on page 254 or by writing them on the chalkboard.

Cover the content of the chapter, beginning with the first section on page 255 and moving through the content. Have students read the material. Write key terms and concepts on the chalkboard. Take a field trip to observe wildlife. This may be to one corner of the school grounds to observe plants, insects, and other species, including spiders and mushrooms. Select one of the species and determine the habitat that it needs. This can involve studying the site, as well as using reference materials.

Many students will enjoy having a wildlife conservation officer serve as a resource person in class. They can discuss the wildlife that are found locally, as well as the laws that apply to hunting and fishing. Some students may be interested in forming a hunter safety education group. In some schools, hunter safety education may be a part of the class instruction.

REVIEW AND EVALUATION

Use the "Reviewing" section at the end of the chapter. Begin by having the students read and orally summarize the "Main Ideas." This can be followed by having them answer the "Questions" orally or in writing. Some teachers prefer to have students prepare written answers as homework and discuss the answers in class the next day.

Evaluation can use the review process, as well as the "Chapter Self-Check", teacher-made tests, and other means. Reteaching will be based on the evaluation.

ADDITIONAL RESOURCES

Many students will like to review the game and fish laws for the state where they live. Information on seasons, bag limits, and licensing is available from the state game and fish agency or wildlife protection department. Brochures, posters, and other items can be used.

ANSWERS

QUESTIONS

1. What is the difference between wildlife and game animals?

Wildlife is all non-domesticated living things. Game animals are wildlife but they are hunted as a sport.

2. Why is some wildlife scarce?

Wildlife may be scarce because it has been over-hunted or because of changes in the environment so that they can no longer live.

3. How do people enjoy wildlife?

People enjoy wildlife as food and sport and by watching it in nature.

4. How does the habitat for aquatic and terrestrial animals differ?

Aquatic animals live in water and terrestrial animals live on the land.

5. What are the major areas of animal wildlife habitat management?

Animal habitat includes food, water, cover, and range and territory.

6. What can we do to provide habitat for plant wildlife?

We can learn what plants need to live and grow. We can set aside places where the plants can grow and avoid killing the plants.

CHAPTER SELF-CHECK

1=G, 2=H, 3=C, 4=B, 5=A, 6=D, 7=E, and 8=F.

Chapter 21

PREVENTING POLLUTION

CHAPTER SUMMARY

Pollution is when materials get into the environment that cause harm. Environmental pollution is when people pollute their surroundings. A pollutant is anything that causes pollution. The polluter-must-pay principle says that the people who pollute the environment must pay to have it cleaned up.

Several kinds of pollution cause damage. Noise pollution is any noise that is damaging or a nuisance to people. Noise is measured in dBA (decibels or sound pressure). Cities may have noise ordinances that set a dBA level of 80 as the maximum. Above 75 dBA, noise may damage the hearing of people.

Water pollution is damaging water so that it can't be used. Potable water is suited for human drinking. Water may be polluted by soil particles, sewage, disease pathogens, and toxic chemicals.

Air pollution results when pollutants get into the air. Both inside and outside air may be polluted.

Soil pollution means that the productivity of the soil has been damaged. Chemicals may be spilled on the soil causing it to not be productive or even making it hazardous to the health of humans and other animals.

Land pollution is when areas are polluted with wastes. Junk cars and old machinery are often seen in pastures and fields or pushed off into streams. Solid wastes are garbage and other materials that people discard. Hazardous wastes are materials that harm the environment.

Thermal pollution is when heat or cold get into the environment. Factories may release heated water or steam into the air. These need to be controlled so that the natural environment is not appreciably changed.

Pollution comes from point and nonpoint sources. Point sources are definite and identifiable. Nonpoint sources are broad and frequently occurring. Examples of nonpoint sources include construction sites, agricultural and forestry practices, surface mining, and urban runoff. People can do a lot to prevent pollution.

INSTRUCTIONAL OBJECTIVES

The objectives for the chapter are presented as questions on page 268 in the text. The instructional objectives are presented below.

Upon completion of chapter 21, the student will be able to:

1. Define pollution and describe why it is a problem.
2. List and explain different kinds of pollution.
3. List the common sources of pollution.
4. Describe ways of preventing pollution.

INTEREST APPROACH, INSTRUCTIONAL STRATEGIES, AND TEACHING PLANS

The overall emphasis with this chapter is that people want to live in a clean environment. They don't want to live in a place that is dirty. To have a clean environment, everyone needs to assume responsibility for keeping the environment clean.

A possible interest approach is to have the students read the introductory part of the chapter. Ask the students to give examples of pollution. Ask, "Have you seen people throw trash out of cars along the roadside? How did you feel about what you saw?"

Present the objectives for the chapter by having students read the questions on page 268 or list the objectives on the chalkboard.

Begin covering the chapter by having the students read the sections of the chapter in sequence. Following each section, discuss the content and write key words and concepts on the chalkboard. Involve the students in summarizing the information. Ask the students their attitudes about the "polluter-must-pay principle." (This should strengthen the affective learning of the students about pollution.)

Summarize each of the kinds of pollution. Have the students name the important concepts associated with each kind. Summarize the key terms on the chalkboard.

After the sources of pollution have been presented, have the students name sources of pollution in their local community. List these on the chalkboard. Have the students suggest or determine solutions to the pollution. Initiate a class project to clean up a polluted area.

REVIEW AND EVALUATION

Review the chapter content by using the "Reviewing" section on page 279. Have students read the "Main Ideas" and orally summarize the content. Also have the students provide answers to the "Questions."

Evaluation should involve the "Chapter Self-Check", teacher-made tests, tests from the computer test bank, or performance on the review.

ADDITIONAL RESOURCES

Brochures and other information about pollution programs in the local community will be useful. An official with a local government agency can probably provide information on local laws related to the different areas of pollution. The class may wish to get information on how to start a community clean-up program, such as adopting a section of road. Contact the state or local highway department for the information.

ANSWERS

QUESTIONS

1. **What is the polluter-must-pay principle? Do you agree with it? Why?**

   The principle says that the person who creates pollution must pay to clean it up. Opinions differ on the principle.

2. **List the six kinds of pollution and examples of each.**

   The six kinds of pollution and examples of each are:

   noise — loud noise causes by airplanes, musical instruments, etc.

   water — damage to water caused by soil particles, sewage, disease pathogens, and chemicals.

   air — different materials get into the air, such as dust, smoke, and oxides.

   soil — when the capacity of the soil to produce is destroyed by chemical spills or other pollutants.

   land — when land is damaged by solid wastes or hazardous wastes.

   thermal — when water or other material is released into the environment so that the natural temperature is changed.

3. **What is sewage? How does if differ from effluent?**

   Sewage is the organic waste made by people. Sewage contains solid materials. Effluent is from manufacturing and contains little solid material.

4. **What is inside and outside air pollution? Why are these important?**

   Inside pollution is inside of buildings, while outside pollution is outside of buildings. Both forms are important because they are hazardous to the health of people and other living things.

5. Why is it important to use biodegradable products?

Biodegradable products decay quickly and don't leave pollution for a long time.

6. What are the major nonpoint pollution sources? List examples of each that you know of in your local community.

The major nonpoint pollution sources and examples are:

construction sites — buildings and highways.

agricultural practices — plowing land, using pesticides, raising animals.

forestry practices — cutting trees and driving vehicles through the forest.

surface mining — moving topsoil to get at minerals, such as minerals and fuel.

urban runoff — asphalt roofs and streets have oil in and on them; it is washed into streams and lakes when it rains.

CHAPTER SELF-CHECK

1=G, 2=H, 3=A, 4=B, 5=C, 6=D, 7=E, and 8=F.

Chapter 22

RECYCLING

CHAPTER SUMMARY

Recycling is reusing something. In many cases, recycling may involve sending the product back to a factory to be made into another product. The U.S. Environmental Protection Agency defines recycling as "collecting, reprocessing, marketing, and using materials once considered trash."

Four major kinds of recycling are used: natural recycling, which is the way nature restores a product to the earth; reuse recycling, which is using something again; remanufacturing recycling, which is making a product into another product; and direct recycling, which occurs when a manufacturer sends a defective product back through the manufacturing process.

Recycling is important because it provides a good use for waste. It helps conserve resources and reduce pollution. Recycling reduces damage to the ozone layer and helps prevent the greenhouse effect. It also saves the cost of moving trash.

Many different products can be recycled. These include water and animal wastes, paper, plastic, metal, glass, oils and paints, and food and feed products.

Citizens often find it relatively easy to recycle. Some communities have curbside pick-up of items to be recycled. Other communities have drop-off centers and buyback centers. Many people can compost materials around their homes.

INSTRUCTIONAL OBJECTIVES

The objectives for the chapter are stated as questions on page 282 of the text. The instructional objectives are listed below.

Upon completion of chapter 22, the student will be able to:

1. Define recycling.
2. Explain why recycling is important.
3. List the types of products that can be recycled.
4. Discuss the roles of people in recycling.

INTEREST APPROACH, INSTRUCTIONAL STRATEGIES, AND TEACHING PLANS

Instruction in this chapter should help develop positive attitudes toward recycling materials. In some cases, the students may wish to be active in initiating a recycling program at the school or in the community.

The interest approach can involve the students reading the introductory part of chapter 22. Ask the students if they have recycled materials. Ask them to indicate what it was and how it was done.

Present the objectives for the chapter by having the students read the questions on page 282 or writing the objective on the chalkboard.

Cover the content of the chapter by having students read the different sections and orally summarize what they have read. Key terms and concepts should be listed on the chalkboard.

Have students prepare a map of the community or general area where they live that identifies where materials can be recycled. This includes drop-off and buyback centers. Of course, many retail food stores will pay a deposit on returnable bottles. Stores in some states will pay a deposit on bottles to be recycled. The class may also plan a school recycling program.

REVIEW AND EVALUATION

Review can involve using the "Reviewing" section at the end of the chapter. Have students read the "Main Ideas" and orally summarize what they have read. Also, have the students use the "Questions" and provide oral or written answers.

Evaluation can involve the review activities, as well as the "Chapter Self-Check" or teacher-made tests.

Reteaching should be based on the review and the evaluation.

ADDITIONAL RESOURCES

Information about recycling programs in the local community can be helpful with this chapter. Many communities have pamphlets or other items that describe how to recycle. The local office of the Cooperative Extension Service will likely have materials that will be useful.

ANSWERS

QUESTIONS

1. Why is recycling better than taking trash to a landfill?

Recycling results in new products. Taking trash to a landfill is expensive and fills up the landfill.

2. Distinguish between the different types of recycling.

The types of recycling are:
natural — the process nature uses to restore the earth.
reuse — reusing something.
remanufacturing — making a product that has been used into a new product.
direct — when a manufacturer sends a defective product back for remanufacturing.

3. Why is recycling important?

Recycling is important because it provides a good use for wastes, conserves resources, reduces pollution, reduces damage to the ozone layer, helps prevent the greenhouse effect, and save the cost of hauling and disposing of trash.

4. What kinds of wastes can be recycled?

Wastes that can be recycled are: water and animal wastes, paper, plastic, metal, glass, oils and paints, and food and feed products.

5. How much waste does a person produce?

It is estimated that a person produces 600 times their weight in waste during the average life time.

6. What is composting? How is it beneficial?

Composting is a natural way of recycling. People set aside places in their yards where food scraps and yard wastes are allowed to decay. Compost saves the cost of disposing of the waste and producing a good quality fertilizer.

CHAPTER SELF-CHECK

1=C, 2=H, 3=D, 4=E, 5=F, 6=G, 7=B, and 8=A.

Chapter 23

CAREERS IN AGRISCIENCE

CHAPTER SUMMARY

People can be successful in a wide range of careers in agriscience. Planning and preparation will be needed. The United States economy is based on people being productive in gainful employment. This is the work ethic of the majority of the people. Work has been viewed as a natural and integral part of life. All able people should work in gainful employment. This means that they work for pay.

An agriscience occupation is any occupation that requires a person to have knowledge in an agricultural area. Many people have a sequence or series of occupations over a long period of time that become a career. Career opportunities are found in many areas of agriscience occupations. Many of the opportunities will be in agribusiness and not on the farm in the production of plants and animals. These occupations are found in all phases of providing products that meet the needs of people. The major areas of occupations are: scientists, engineers, and related professionals; marketing, merchandising, and sales representatives; managers and financial specialists; social service professionals; education and communication; and production agriculture.

INSTRUCTIONAL OBJECTIVES

The objectives for the chapter are stated as questions on page 296 of the text. The instructional objectives are presented here.

Upon completion of chapter 23, the student will be able to:

1. Define the terms occupation, job, and career.

2. Describe the area of agriscience as a career.

3. List different types of agriscience occupations.

4. Describe the nature of the work in selected agriscience occupations.

5. List the factors to consider when deciding about an agriscience occupation.

INTEREST APPROACH, INSTRUCTIONAL STRATEGIES, AND TEACHING PLANS

Teachers often need more creativity in teaching about careers and employment than in other areas of agriscience. Students sometimes view the study of careers as boring. To help overcome this problem, a "Career Profile" has been included in each chapter.

The interest approach can involve having the students read the introductory section of the chapter. Afterward, ask students why people work and have them name examples of work in the local community.

Present the objectives using the questions on page 296 or by writing the objectives on the chalkboard.

An appropriate strategy with the content of the chapter is to have students read each section and follow up the reading with discussion. List key terms and concepts on the chalkboard. Review the lists and examples presented in the text.

Students will enjoy field trips to observe people in agriscience occupations or having resource people come to class to tell about their work. In some cases, the students may want to job shadow an agriscience worker. Of course, all details of this should be in line with school procedures.

The students may each be assigned an agriscience occupation to research and report in writing or orally to the class. Providing a camera to the students, to make photographs of people at work in the occupation they select, to prepare a poster would give added meaning to this activity. (Cameras for this kind of activity may be provided by the Polaroid Corporation of Cambridge, Massachusetts. The cameras are sold to teachers and schools at reduced prices through a special Visual Learning Program.)

The *AgriScience Interest Inventory* (Interstate Publishers, Inc.) will be a good computer-based tool to use with this chapter.

REVIEW AND EVALUATION

Review can involve using the "Main Ideas" and "Questions" parts of the end of chapter "Reviewing" section. Have students read the "Main Ideas" and provide an oral summary to the class. Have students prepare answers to the questions and orally report the answers to the class.

Evaluation can involve the review activities, as well as the "Chapter Self-Check." The "Agriscience Vocabulary" list on page 296 may also be helpful in evaluation.

ADDITIONAL RESOURCES

Additional resources can be particularly useful with this chapter. Brochures and occupational guidance materials for agriculture may be useful.

A good book to use is *Careers in Agribusiness and Industry* (Interstate Publishers, Inc.). Students can review this book for information about agriscience occupations.

ANSWERS

QUESTIONS

1. **Contrast jobs, occupations, and careers.**

   A job is specific work that involves a specific site and employer. An occupation is work that can be given a title. A career is a series of jobs over a long time.

2. **How does an agriscience job differ from a job that is not an agriscience job?**

   An agriscience job is any job that requires the person to have some knowledge of agriculture.

3. **List the six different types of agriscience jobs?**

   The six different types of agriscience jobs are:

   (1) marketing, merchandising, and sales representatives;

   (2) scientists, engineers, and related professionals;

(3) managers and financial specialists;
(4) social service professionals;
(5) education and communication; and
(6) production.

4. For each type of agriscience job listed in question 3, give three examples of specific jobs in each category.

(The types of jobs for the six types are listed in the text on pages 301-303. Refer to these lists for the jobs.)

5. If a person came to you and indicated an interest in an agriscience job, what are the things that person should consider before taking that job?

Three things to consider are: your likes, education needed, and opportunities for advancement.

CHAPTER SELF-CHECK

1=F, 2=E, 3=G, 4=C, 5=B, 6=A, 7=H and 8=D.

Chapter 24

GETTING THE EDUCATION

CHAPTER SUMMARY

Success in a career requires the appropriate education. Two kinds of education are needed: informal and formal. Informal education is the education people get from their environment. Formal education is the education people get from classes and going to school. Education requirements vary; people must have education that is commensurate with the occupations they want to have. Success in getting a job requires the appropriate education, experience, and personal habits

Students need to take the right courses in high school so that they can go on for additional education or enter the occupation of their choice. Vocational education can develop many needed occupational skills. This is so for those who enter work after high school and for those who continue education in institutes or colleges. A minimum at some colleges for admission is 4 units of English, 2 of algebra and geometry, 2 of science (biology, chemistry, and physics), 2 of social studies, and 3 of foreign language. Other colleges require more science and mathematics and less foreign language.

INSTRUCTIONAL OBJECTIVES

The objectives for the chapter are presented as questions on page 308. Teachers may choose to use the questions or the objectives listed below.

Upon completion of chapter 24, the student will be able to:

1. List the education needed for some agriscience occupations.

2. List examples of salaries earned in agriscience jobs.

3. Explain what is needed to be successful in getting an agriscience job.
4. List examples of courses to take in high school for college admission.
5. List sources of information about getting the education.

INTEREST APPROACH, INSTRUCTIONAL STRATEGIES, AND TEACHING PLANS

A variety of procedures should be used with this chapter. An admissions counselor from a college with an agriculture program could be used as a resource person in class. The focus should be on gaining admission and being successful in studying for an agriscience career.

Have students read the introductory section of the chapter. Ask them to orally summarize what they have read. Ask them if they have thought about the education they need to enter a career. See if any will volunteer their plans.

Present the objectives for the chapter by referring the students to the questions on page 308 or listing the objectives on the chalkboard.

Have students read the sections of the chapter. Ask them to orally report what they have read. Summarize key terms and concepts on the chalkboard. Students may wish to tour the agricultural education facilities of the local high school and learn about the classes that are offered. Get admissions information from a college and have students review the requirements. People in various agriscience careers may serve as resource persons in the class.

REVIEW AND EVALUATION

The chapter can be reviewed by using the "Reviewing" section, including the "Main Ideas" and "Questions." Have students read and orally summarize the "Main Ideas." Students can answer the questions in writing for homework or during supervised study or orally during class.

Evaluation can involve using the "Chapter Self-Check" and other devices that are prepared by the teacher.

Reteaching can be based on observations of the students during the review and evaluation.

ADDITIONAL RESOURCES

Several additional resources are included on pages 316 and 317 of the text. These may be available in the school counselor's office, library, or ordered for the agriculture department. Catalogs and admissions information from various schools will be helpful.

ANSWERS

QUESTIONS

1. **Describe in your own words informal education and formal education. Where could each type of education be pursued in your community?**

   The acceptable answer should indicate that informal education is the education people get from their environment and formal education is what people get in classes at school. Where they are pursued locally varies with the community.

2. **List in sequential order the levels of education students would go through from the first time they enter school until they complete the highest level.**

   The levels are kindergarten, elementary school, middle or junior high school, senior high school, and postsecondary education, including colleges and universities.

3. **What are the four different levels of jobs a person might enter in the labor market?**

   The levels of jobs are unskilled, skilled, semiprofessional, and professional.

4. **What does an employer consider or review about an individual when hiring a person to fill a job?**

   An employer considers education, experience, and personal habits.

5. **List the subjects many colleges or universities recommend that students take who**

are interested in pursuing postsecondary education.

The subjects students often need for admission are: English (4 units), algebra and geometry (2 units), science (2 units), history and social studies (2 units), and foreign language (3 units).

CHAPTER SELF-CHECK

1=F, 2=D, 3=E, 4=C, 5=B, 6=G, 7=A, and 8=H.

Chapter 25

DEVELOPING PERSONAL SKILLS

CHAPTER SUMMARY

Successful people get along well with other people. The traits that people have that help them get along with other people are known as personal skills. People can develop these skills and should set about to deliberately do so in order to be productive in their work.

People need two major kinds of personal skills to be successful. Work skills are the skills that people need to get, hold, and advance in their work. Interpersonal skills are the skills that people have in relating to other people, and are very important in career success.

Human relations deals with how people relate to each other. Good human relations comes about when we try to understand other people and relate to them.

Communications is an important area of personal skill. Communication is exchanging information. People have to select the medium to exchange the information. This requires knowing something about the people we wish to communicate with. The medium must be used properly so that communication occurs.

Leadership skills are important personal skills. Leadership is the ability to influence other people to achieve certain goals or activities. The role of the leader varies with the situation, known as situational leadership. Effective leaders tend to have certain traits that help them to be successful.

Personal appearance is a personal skill that can be used to advantage. Some people present themselves very poorly. Grooming and clothing are two areas of personal appearance that can be used to help people be successful. People need to remember that it is important to present themselves in a positive way. Certain grooming and clothing is valued in our culture, and successful people will do what is expected.

INSTRUCTIONAL OBJECTIVES

The objectives for the chapter are presented below and in the text on page 320 as questions.

Upon completion of chapter 25, the student will be able to:

1. Explain important personal skills.
2. Describe the meaning and importance of human relations.
3. Explain why communication is important.

4. Define leadership and list qualities of leaders.
5. Explain why dress and grooming are important and list acceptable practices.

INTEREST APPROACH, INSTRUCTIONAL STRATEGIES, AND TEACHING PLANS

This chapter could be interesting and controversial. Students often don't understand the importance of grooming and dress. They sometimes indicate that they don't want to conform to what is traditionally expected. Teachers will want to use the differences of opinion as instructional tools in teaching the chapter.

Begin the interest approach by having students read the introductory section on page 319. Ask them what it means when "people get along well with other people." Get the students to give examples.

Present the objectives for the chapter by using the questions on page 320 or writing the objectives on the chalkboard.

Cover the content of the chapter by having students read the sections of the chapter. The sequence of the content could be varied if it is desired. Student discussion and summarizing key terms and concepts on the chalkboard will be important.

Role playing can be used in teaching the chapter. Students can role play good and poor approaches to handling human relations and communication.

Using resource people to discuss appropriate grooming and dress would be useful. A demonstration or style show would be of high interest to the students.

This chapter is a good time for students to learn about the role of student organizations in teaching leadership skills. FFA officers or leaders in other organizations can present a demonstration in class on ways of introducing people or conducting a meeting.

REVIEW AND EVALUATION

Use the "Reviewing" section at the end of the chapter. Have students read "Main Ideas" and orally summarize the content. The Questions could be answered in writing as homework or during supervised study or answered orally in class.

Evaluation can involve observing the performance of the students on the review, as well as on the "Chapter Self-Check". Some teachers may prepare other tests for the class. The computer test bank will be helpful.

Reteaching should be based on observations during review and evaluation.

ADDITIONAL RESOURCES

The additional resources for this chapter will likely focus more on people than materials. Resource persons on grooming, dress, leadership, communication, and other personal skills will be helpful. The Junior ROTC program in the high school may have useful materials on personal skill development.

ANSWERS

QUESTIONS

1. **What are the four areas of personal skills? Briefly explain each.**

   The four areas of personal skill are:
   (1) work skills — skills that people have to help them get and hold a job.
   (2) citizenship skills — skills that help people be good members of their communities.
   (3) living skills — skills in how to carry on the routine of life.
   (4) interpersonal skills — skills in relating to other people.

2. **What is human relations? Why is career success tied to good human relations?**

   Human relations deals with how people relate to each other. Career success involves getting along with people.

3. What is involved in good communications?

Communication is the exchange of information. People must select the medium, know the other person, and use the medium properly.

4. What are the characteristics of effective leaders?

Effective leaders are good in communications, builders of strong groups, and honest and fair. They help other people develop skills and hold high expectations for them. Leaders set good examples and enjoy life.

5. Why is personal appearance important?

Personal appearance is associated with being successful. People feel better about themselves if they present a good appearance.

6. Briefly explain two important areas in personal appearance.

The two areas are grooming and clothing. Grooming refers to a neat and tidy personal appearance. Clothing is also known as dress, which is what people wear and how they wear it.

CHAPTER SELF-CHECK

1=B, 2=H, 3=C, 4=E, 5=G, 6=D, 7=A, and 8=F.

Chapter 26

BEING A GOOD CITIZEN

CHAPTER SUMMARY

Citizens are people who live in particular places. People tend to think of citizens as individuals who hold citizenship in the United States or other nation. But, not all people who live in a place hold citizenship there. Nationality means that a person belongs to a particular nation. Aliens are people who are living in one country but hold citizenship in another.

In addition to its legal meaning, citizenship means that an individual has certain rights, duties, and responsibilities. A right is something that is due to people. Rights may be granted by law or gained in other ways. The Bill of Rights in the United States listed 10 freedoms. Other freedoms have been added through amendments to the U.S. Constitution. Duties are the things that citizens must do, such as obey the laws and pay taxes. Responsibilities are the things that a person should do. Citizenship education for young people usually includes rights, duties, and responsibilities.

Good citizenship requires certain things of people. The United States has provided a modern, comfortable place to live for most citizens. Youth need to learn the skills to continue a vibrant nation. A few areas where the benefits of practicing good citizenship are important are community pride, healthy environment, educational opportunities, and economic opportunities.

INSTRUCTIONAL OBJECTIVES

The student objectives for this chapter are presented as questions on page 332. The instructional objectives are presented below.

Upon completion of chapter 26, the student will be able to:

1. Describe the meaning of citizenship.
2. List the duties and responsibilities of citizens.
3. Explain the benefits of being a good citizen.

INTEREST APPROACH, INSTRUCTIONAL STRATEGIES, AND TEACHING PLANS

The chapter is intended to help students understand their roles as citizens. The instructional process may involve studying local government and inviting local government officials to serve as resource persons. A field trip to the city hall or courthouse may be useful. Helping students understand the importance of being an informed voter is also important.

Begin the interest approach by having students read the introductory section of the chapter on page 331. Ask if they know of a section of road that has been adopted. Perhaps there is interest in the class adopting a section of road through the school FFA chapter or other organization.

Present the objectives for the chapter by having students read the questions on page 332 or writing the objectives on the chalkboard.

Cover the content of the chapter by having students read the chapter as homework or during supervised study. Go over each section by discussing it in class and writing key terms and concepts on the chalkboard. Particularly emphasize the duties and responsibilities of citizens. Help students explore the benefits of citizen involvement in their local community. Do they know of civic clubs that have taken on community projects? How did these make the community a better place?

REVIEW AND EVALUATION

Review and evaluation often involve some of the same procedures. All should focus on the achievement of the objectives for the learning.

Use the "Main Ideas" and "Questions" in the "Reviewing" section at the end of the chapter. Have students read the "Main Ideas" and provide oral summaries in class. Students may answer the "Questions" as homework or during supervised study or orally during the class.

Evaluation can involve the "Chapter Self-Check" and other means, such as a test constructed from the computer test bank. Teachers often make their own tests to help localize the evaluation to the needs of the students and situations in the community.

Review, evaluation, and reteaching can also be done through activities of student organizations in the school.

ADDITIONAL RESOURCES

Any information about the local government will be helpful. Lists of local elected officials, election dates, taxation, ordinances, and meetings of boards will be useful. Contact the offices of the mayor, county administrator, or other official.

ANSWERS

QUESTIONS

1. **Distinguish between being a citizen and citizenship.**

    A citizen is a person who lives in a particular place. Citizenship means that the citizen has certain rights, duties, and responsibilities.

2. **Why is the right to start and own a business important in agriculture?**

    People can have farms and agribusinesses. When a people own something they take more interest in it. This results in people working to provide a good product and make a profit.

3. **What are the major duties of United States citizens?**

The major duties of United States citizens are: obey the laws, pay taxes, serve as jury members, testify in court, serve in the armed forces, and go to school.

4. What are the major responsibilities of United States citizens?

The major responsibilities of citizens are: vote, take an interest in government, serve in government positions (if elected), help enforce the laws, be honest and fair, respect government officials, respect the personal worth of other people, and help conserve natural resources.

5. What are the major benefits of good citizenship?

The benefits of good citizenship are: community pride, healthy environment, educational opportunities, and economic opportunities.

CHAPTER SELF-CHECK

1=G, 2=H, 3=B, 4=C, 5=A, 6=D, 7=E, and 8=F.

Chapter 27

KEEPING HEALTHY

CHAPTER SUMMARY

Health is the absence of disease. It is the physical and mental condition of a person. Healthy people feel good, have a good outlook on life, and get along well with other people.

Life expectancy is the number of years that a person is expected to live. The life expectancy of women is slightly over seven years more than men. The average life expectancy of all people is 74.7 years. Average life expectancy is going up each year. Increasing the number of years people live involves preventing and controlling disease. Physical, mental, and social health are all important.

Many people follow wellness programs in their effort to have good health. Wellness is the process of having and maintaining physical health, as well as mental and social health. Life style is an important part of wellness. Life style is the way people go about life. Research has shown that the following six practices add years to a person's life:

- Regular physical activity.

- Not smoking.

- Getting seven to eight hours of sleep each night.

- Maintaining the right weight.

- Eating breakfast and other nutritionally sound meals.

- Avoiding drugs and alcohol.

Some people practice fitness as a part of wellness. Fitness is the condition of the human body. Regular exercise is a major part of fitness.

INSTRUCTIONAL OBJECTIVES

The objectives for the chapter are presented as questions on page 342 in the text. The instructional objectives are presented below.

Upon completion of chapter 27, the student will be able to:

1. Define health.
2. Explain the importance of good physical, mental, and social health.
3. Define life style and list desirable life style practices.
4. Describe wellness.

INTEREST APPROACH, INSTRUCTIONAL STRATEGIES, AND TEACHING PLANS

Some students are much involved with wellness and maintaining good health. Others may be doing very little to keep themselves in good condition. The teaching process must not embarrass any student who is over weight or otherwise in poor health. In some cases, the assistance of local drug education officials may be appropriate.

For the interest approach, have the students read the introductory section of the chapter. Ask them to explain what "feeling great and looking good" mean to them. Explain that all of us can feel great and look good if we pay attention to our every day lives.

Present the objectives for the chapter by having the students read the questions on page 342 or writing the objectives on the chalkboard.

Cover the content of the chapter one section at a time. Have students read the sections and discuss them in class. Write key terms and concepts on the chalkboard. Use several resource persons with the chapter. Use the school nurse or other health official to cover physical health, as appropriate. Use the school psychologist or other appropriate person to cover mental health, as appropriate. In some schools, sex education classes may be offered. In other schools, this chapter may be an opportunity for the school nurse or other official to cover this area. (Be sure that all instruction is in line with the policies of the school.) A wellness specialist may be used to help student design fitness and wellness programs for themselves. A field trip to a wellness center may be appropriate.

REVIEW AND EVALUATION

Use the "Reviewing" section at the end of the chapter to help with review and evaluation. Have student read the "Main Ideas" and answer the questions. Written and oral review will be beneficial.

Evaluation can involve the "Chapter Self-Check" and other procedures, such as teacher-made tests and how the students improve their wellness practices.

ADDITIONAL RESOURCES

Brochures on health, wellness, sex education, eating, and other areas may be appropriate. These are available from the local health department, a hospital, the local office of the Cooperative Extension Service, and other sources.

ANSWERS

QUESTIONS

1. **What is health? How is it related to life expectancy?**

    Health is the physical or mental condition of a person. Having good health increases life expectancy.

2. **What five areas are important for physical health?**

    The five important areas for physical health are: eating right, getting enough sleep and rest, getting exercise, keeping clean, and getting proper medical care.

3. **What problems are associated with mental health?**

    Mental health problems include anxiety, self-esteem, depression, teenage pregnancy, eating disorders, drugs and alcohol, violence, and grief.

4. How are physical and mental health related?

People who strive to have good physical health are more likely to have good mental health. The practices needed for physical health are good for mental health.

5. What is social health?

Social health is how people relate to other people. It involves communicating and sharing life.

6. What life style practices are important?

Life study is the way people go about living. Several practices are important in life style: exercise, not smoking, adequate sleep, the right weight, eating breakfast and other nutritionally sound meals, and not using drugs and alcohol.

CHAPTER SELF-CHECK

1=H, 2=A, 3=B, 4=C, 5=E, 6=F, 7=D, and 8=G.